STEAM Education

STEAM Education

Intersections and Thresholds

Edited by

Yichien Cooper and Alice Lai

BRILL

LEIDEN | BOSTON

Cover illustration: Design by Emilie Cooper

All chapters in this book have undergone peer review.

The Library of Congress Cataloging-in-Publication Data is available online at https://catalog.loc.gov

Typeface for the Latin, Greek, and Cyrillic scripts: "Brill". See and download: brill.com/brill-typeface.

ISBN 978-90-04-71472-4 (paperback)
ISBN 978-90-04-71473-1 (hardback)
ISBN 978-90-04-71474-8 (e-book)
DOI 10.1163/9789004714748

This book is printed on acid-free paper and produced in a sustainable manner.

Contents

Acknowledgments IX
List of Figures and Tables X
Notes on Contributors XIV

Introduction: Celebrating Diverse Perspectives on STEAM Education 1
Yichien Cooper and Alice Lai

PART 1
Rethinking STEAM Frameworks

1 You Don't Need a List of the Best STEAM Lesson Ideas: Focusing on the Transdisciplinary Competencies 21
Christine Liao

2 Cultivating Authentic Approaches to STEAM Teaching and Learning through the Arts and Design 38
Debrah C. Sickler-Voigt

3 Challenges and Prospects for STEAM Education in Japan: Considerations from the Perspective of Art Education 51
Isao Ohashi

4 Effect of STEAM Education on Schools and Universities in Türkiye 69
Vedat Özsoy

PART 2
Redirecting STEAM Foci in the Communities

5 Reflections on the Practice of STEAM Education in Penang, Malaysia 85
Paul Boey

6 Natural Aesthetics: Interdisciplinary Learning of the Arts and Natural Science 97
Chia-Hung Kao

7 Reflections on the Design of a STEAM Curriculum for 12- to 14-Year-Old Girls 107
Eliza Pitri, Maria Evagorou and Agni Stylianou-Georgiou

8 Application of Data Visualization in Hakka Cultural Education in Taiwan 126
Shyh-Shiun Shyu and Yui-Chih Wu

9 Promoting DEI with STEAM Technology for People with Visual Impairment and Blindness in Taiwan 145
Hsin-Yi Chao

PART 3
Reimagining STEM as STEAM

10 "Uncharted Territory": Interactive Augmented Reality for Exploring Concepts in Chemistry with Science-Averse College Students 161
Jonah B. Firestone and Don McMahon

11 Incorporating Art in Mathematics Teacher Education: Examples with Primes and Pendulums 181
David Glassmeyer and Kevin Hsieh

12 STEAM in Higher Education Art Programs 193
Snow Yunxue Fu and Luke Hampton

13 Expressive STEM Storymaking: Art, Literacy, and Creative Computing 210
Sean Justice and Lori Czop Assaf

14 Considering STEAM Aesthetics through an Arts-Driven STEAM Pedagogy 227
Alice Lai

PART 4
Remapping Teacher Education through STEAM Practices

15 STEAM-Oriented Instructional Design in Visual Arts Teacher Education in Türkiye 249
Suzan Duygu Bedir Erişti and Yeliz Erdoğan

16 Teaching STEAM Education in Hong Kong 264
So Lan Wong

17 Developing Visual Art Workshops for STEM Teachers 281
Kevin Hsieh and Michael White

Index 295

Acknowledgments

We (Yichien and Alice) deeply appreciate our publisher and editorial team, particularly John Bennet and Christine Hededam at Brill, who have provided us with clear guidance throughout the process of producing this collection. Its completion would have been impossible without the trust and support of Dr. Patricia Leavy, who invited Yichien to apply her expertise in arts integration to contribute to the Art Plus series at Brill. Thanks to her for her generosity in providing vital suggestions during the book proposal process. We are profoundly grateful for our amazing copyeditor Dr. Linda Meixner for her invaluable, meticulous, and plentiful comments, ensuring the stylistic features of this book. We are indebted to Dr. Debrah Sickler-Voigt for inviting us to present early content at the 2023 National Art Education Association convention and to the NAEA Higher Education Forum for the opportunity to share a snippet of the introductory chapter at the 2024 NAEA convention. We acknowledge the timely feedback of Dr. Enid Zimmerman, pointing out the main attractions of this book. Thanks to all our contributors from around the world for their patience and cooperation in making this anthology possible and most importantly for graciously working with us to share their insights on STEAM education despite linguistic and time-zone constraints.

Yichien is thankful for the opportunity to expand the meaning and practices of integrating the arts with support from Dr. Judith Morrison, academic director of the College of Education at Washington State University, Tri-Cities. In addition, she appreciates insightful conversations she had with her preservice teachers at WSUTC, workshop attendees around the world, and young students in local communities. She is grateful to her husband Matt and her children Emilie, Katherine, and Benjamin for their unflagging encouragement and dedicates this book to her parents.

Alice acknowledges the professional development funds provided by Empire State University, State University of New York, and the earnest support of Dr. Nicola Allain, dean of the School of Arts and Humanities. She also extends her gratitude to her family, especially Dr. Eric Ball, for engaging in insightful and critical dialogue about ever-changing educational contexts in a global society.

Figures and Tables

Figures

I.1 STEAM concept expanding IPOP theory (Perkarik et al., 2014) and curricula as bridges (Cooper, 2013). 5
I.2 Pinhole photography, 2021. Exposure from January 10 to June 10, 2021. 6
I.3 Example of tensegrity sculpture by Emilie Cooper, 2020. 7
1.1 The relationship between STEAM teaching approaches, STEAM competencies, and SEL skills. 28
2.1 This STEAM logo includes visual icons representing the content of *STEAM Teaching and Learning Through the Arts and Design: A Practical Guide for PK–12 Educators* (Sickler-Voigt, 2023). Its design symbolically unifies the STEAM disciplines and represents the value of interdisciplinary and transdisciplinary studies. © Debrah Sickler-Voigt. 46
2.2 This photo of the tunnel book shown with white LED lighting showcases its removable hand-assembled rabbit ears. © Debrah Sickler-Voigt, Paige Brenner, and Richard Sickler. 47
2.3 This photo collage shows the tunnel book with its range of LED hues. Its associated video illustrates its dynamic multimedia qualities (see Sickler & Sickler-Voigt, 2023). © Debrah Sickler-Voigt, Paige Brenner, and Richard Sickler. 48
3.1 Learning structure as a process of self-actualization based on Ohashi (2021) and updated for this chapter. 58
3.2 Rabbits with a sword and a fishing rod. 60
3.3 Out! The crocodiles are coming! (If you fail in a game programmed with the Micro:bit, you will be attacked by an alligator.). 61
3.4 The children used the PC as fluently as they used scissors and glue to give shape to their ideas. 62
3.5 Paper Cup Tower Activity, conceived by Dr. Tetsuo Kiyota and practiced by Ms. Hiroko Hanazato at Nakano 5th Junior High School, Tokyo. 66
4.1 STEAM training examples: weather. 71
4.2 STEAM education applications of a Grade 7 student, 2022. 76
5.1 I used scrap tires to create an installation art, depicting an endangered dusky leaf monkey in Myanmar, Malaysia. 91
5.2 For the *Heart of the Ocean* community art education exhibition, I collaborated with teachers and students to create an installation piece on campus depicting a stranded whale. The purpose of this collaborative effort

was to inspire others and raise awareness about important topics related to environmental protection. 92
6.1 Emblem of the National Museum of Natural Science. 98
6.2 A conceptual map of the pedagogical content for the natural aesthetics curriculum that aligns with the three core aspects of aesthetics education in Taiwan. 99
6.3 The spotted, strap-shaped, and thematic learning approaches. 100
6.4 A learner sew plant specimens. 103
6.5 Learner made a piece of paper in the shape of Taiwan by using plant fibers and stencils. 104
6.6 Student's example of an island feast. 106
7.1 Photo from Day 4: final product created by the groups (uniforms for four people working in STEM). 114
7.2 Photo from Day 1: exploring a lab in a virtual environment. 117
7.3 Photo from Day 2: observing organisms collected from the soil at the park. 117
7.4 Photo from Day 5: working on the prototype of the product, a pillow that can change color. 118
8.1 Teaching method for online summer camp curriculum. 132
8.2 Left to right: participant L's and K's Data Visualization Project from Unit 1. 135
8.3 Left to right: projects of participant C and D from Unit 2. 137
8.4 Left to right: participant J's and D's data visualization project from Unit 3. 141
9.1 Six capabilities of visual art education for students with visual impairments and blindness. 150
9.2 Functional flow chart of National Taiwan Museum of Fine Arts Accessible Application for the visual and hearing-impaired audience features three guide models: nearby, map, and list with extended usage. Created by Hsin-Yi Chao. 152
9.3 Universal color wheel with visual and tangible textures was invented by Chao in 2022. 156
10.1 The reality–virtuality continuum (Milgram et al., 1994, p. 283). 162
11.1 The Desmos graph of class data (green) with previous class data (blue), comparing the length of the string on the horizontal x-axis (measured in meters) to the period (measured in seconds). 187
11.2 The Desmos graph of a function modeling the relationship between the pendulum string length and period. 188

12.1 Screenshot of an avatar running through an abstracted city scape. Fu, Y. (n.d.). *Run – Valley City Arts Game Recording Building Level (Take2), 2021* (from https://snowyunxuefu.com/section/505997-Run%20-%20Valley%20City.html). 199
12.2 Virtual aerial view of VR WSPark. Fu, Y. (n.d.). *VR WSPark metaverse project* (from https://snowyunxuefu.com/artwork/5087411ArtLabb%20Opening%20on%20VR%20WSPark.html). 204
12.3 YouTube thumbnail advertising the online Metaverse Artlabb Show Opening. Fu, Y. (n.d.). *VR WSPark metaverse project* (from https://snowyunxuefu.com/artwork/5087411ArtLabb%20Opening%20on%20VR%20WSPark.html). 204
13.1 Camouflaged sight words in ScratchJr. The right screen shows a sight word hiding in the trees. The left screen shows the student's code: when the cat is touched, ScratchJr. will play audio made by the student of herself speaking the sight words she has hidden. 216
13.2 Interactive, screen-based storymaking. *Knights of the Woods* (Picou, 2021) combined hand-drawn characters with Scratch's stock characters and dialog, music, and creative coding to advance the narrative. To amplify reader interaction, Picou programmed keystrokes that animate the characters when tapped by a reader. 218
13.3 Transdigital storymaking. *If You Are Given a Cookie …* by Lyla Guidry (2022) is constructed from a paper collage connected to a Scratch animation with a Makey Makey®. Each touch of the paper cookie takes another bite from the screen cookie, producing a munching sound. 220
14.1 Student A's golden rectangles and spirals (left). The predominant geometric shapes placed over the golden rectangles (right), 2022. 237
14.2 Student A's poster design in progress (left). Student A's final poster design (right), 2022. 238
14.3 Student B's sketch of a Fibonacci spiral. 239
14.4 Student B. *Fibonacci Koi Pond.* 8" × 13." Gouache on Stonehenge Print Paper. 2023. 240
14.5 Student C. *Transformation Zone.* Watercolor, 2023. 240
15.1 Kaleidocycle designs of the lecturers and teacher candidates (left to right): Top row: Yeliz Erdoğan, Yeliz Erdoğan, Zehra Dede. Middle row: Esra Kuş, Gamze Gezer, Sena Sevinç. Bottom row: Sümeyye Güneş, Ela Kantarcı, Birsen Çelik, 2020, Eskişehir, Türkiye. 255
15.2 Model designs by Buse Buket Turan, Ela Kantarcı, Sena Sevinç, Esra Kuş (left to right), 2020, Eskişehir, Türkiye. 257
15.3 Kitchen lithography: printmaking with drawing on aluminum foil, emulsion by Buse Buket Turan and Sena Sevinç, 2020, Eskişehir, Türkiye. 258

16.1 Learning activities based on KLA topics. 266
16.2 Learning activities based on projects. 266
16.3 Front (L) and inner (R) pages of a light-up greeting card. 271
16.4 Students' drawings and exploration of various methods to use the Makey Makey® Kit and closed circuits 272
16.5 Using the Makey Makey® Kit and drawings to play a song. 273
16.6 Using reassembled hexagons to create a new tessellation pattern. 275
17.1 In the Making Sketchbooks Workshop, 2020, the instructor gave a step-by-step demonstration on how to use recycled cereal boxes and copy paper to make sketchbooks. 284
17.2 Drawing and painting lecturer showed the processes of 2D design and making patterns. 285
17.3 In the Sculptural Hats Workshop (2021) taught by an art education professor, the participants showed their three-dimensional paper hats. 287

Tables

I.1 Key initiatives advocating for STEAM education nationwide and in the field of art education. 8
I.2 Preliminary data on STEAM-centered articles appearing in selected journals in the field of art education. 9
I.3 Various approaches applied in STEAM education. 10
7.1 STEAM curriculum, Version 1. 112
7.2 STEAM curriculum, Version 2. 115
9.1 The methods of visual art exhibition designed for the blind. 153
10.1 Overall change per question and question type from pretest to posttest. 172

Notes on Contributors

Lori Czop Assaf
is a professor in reading education in the Department of Curriculum and Instruction at Texas State University. She is past director of the Central Texas Writing Project and teaches the study abroad program to South Africa. Her research focuses on teacher learning, teacher identity, and supporting K–12 English Language Learners as readers and writers. Dr. Assaf is committed to international collaboration specifically on culturally mediated writing instruction across the disciplines and professional development for preservice and in-service teachers.

Paul Boey
is a postdoctoral fellow in art from Southeast University in China. His research specialty and interests include Chinese and foreign art history, art education, and the development of culture and art through One Belt One Road in China and Southeast Asia. He served as the director and master's degree supervisor of the Oriental Culture and Art Research Centre of the University of the Thai Chamber of Commerce in 2018 and was appointed visiting professor at Chongqing Technology and Business University in China in 2019. Currently, he is a supervisor of master's and doctoral students at New Era University College in Malaysia and a committee member of the World Chinese Arts Education Association. Boey has presented research-based papers at national and international academic conferences and has authored art-related publications, including two volumes of *A History of Malaysian Art* (2016).

Hsin-Yi Chao
is an assistant professor in the Cultural and Creative Industry program at the National Chung Hsing University in Taiwan. After earning a Master of Fine Arts at the National Taipei University of Education and a Ph.D. in architecture at the National Taiwan University of Science and Technology, she continued postdoctoral research in the psychology graduate program at the University of Toronto in Canada. She spent 10 years as a visual art teacher at a primary school and the last six as a curator working on accessible technology and multisensory exhibitions at the National Taiwan Museum of Fine Arts. The founder of the national nonprofit Taiwan Art Beyond Vision Association, Dr. Chao focuses her research on nonvisual art educational theory and practice, accessible technology applications, universal display design, audio description of art, and tactile–spatial cognition research focusing on special education,

psychology, museums, and art to promote equal rights to art education for those with visual impairment.

Yichien Cooper
is an artist, educator, author, and arts advocate, serving as a career-track assistant professor and field supervisor at Washington State University, Tri-Cities. In her research she explores STEAM education, arts-based pedagogy, data visualization, socially responsible teaching, mixed identity, and Asian aesthetics. Dr. Cooper has not only delivered keynote addresses and conducted workshops in Australia, China, Malaysia, South Korea, Taiwan, Thailand, Türkiye, and the United States, but she has also published extensively on integrating arts curriculum for Chinese educators. As the current Director of the Higher Education Division for the National Art Education Association, she has received numerous honors, including the 2019 City of Richland Proclamation of Appreciation, the 2023 Ziegfeld Service Award from the United States Society of Education through Art, the 2024 Kathy Connors Teaching Award from the NAEA Coalition for Feminisms in Art Education, and the 2024 Washington State University, Tri-Cities Campus Distinguished Teaching Award.

Yeliz Erdoğan
is a native Trabzon, Türkiye, graduated in 2014 from the Ondokuz Mayıs University (OMU) Faculty of Education, specializing in fine art education. Having earned her master's degree in Fine Art Education from OMU in 2018 and her doctorate from Anadolu University Institute of Educational Sciences in 2023 with a concentration on STEAM education, she currently serves as a lecturer at Van Yüzüncü Yıl University, specializing in fine arts education with a focus on STEAM integration. She continues to advance her work in this field while teaching courses on STEAM applications in the visual arts.

Suzan Duygu Bedir Erişti
is an esteemed professor at Anadolu University's Faculty of Education, specializing in fine arts education. Dr. Erişti's academic journey includes master's and doctoral degrees in art, visual design, art history, and education. She is dedicated to innovative pedagogical methods, using art-based and visual research techniques, and pioneering a/r/tography methodologies. Her research interests encompass digital visual culture, AI in art education, technology-driven pedagogy, multicultural art education, graphic design, interactive instructional design, gamification, digital storytelling, and STEAM applications. She explores the intersection of digital visual culture and artificial intelligence in her work. Dr. Erişti is a fervent advocate for international collaboration, participating in

national and international exhibitions, research initiatives, and projects. Recognized as Visual Arts Educator of the Year in 2016, she continues to make significant contributions to academia, demonstrating unwavering dedication to excellence and profound impact in her field.

Maria Evagorou
is an associate professor of science education at the University of Nicosia, Cyprus. Her research focuses on STEM education with an emphasis on integrated STEM and teacher professional development. Dr. Evagorou participates in funded projects on argumentation, socioscientific issues, the role of language and culture in the teaching of STEM [IncluSME], and support for girls who wish to pursue STEM careers. Maria currently leads a policy initiative as part of a European project, placing emphasis on the development of STEM training for in-service and preservice teachers.

Jonah B. Firestone
is currently an associate professor in science education in the College of Education at Washington State University, Tri-Cities. Dr. Firestone has also worked as a science and mathematics teacher for diverse groups of students in Arizona. With over 20 years' experience working with large-scale and high-stakes test preparation, creation, teacher-training, and evaluation in the U.S. and Hong Kong, he is the principal investigator for the Simulation and Integrated Media for Instruction, Assessment, and Neurocognition (SIMIAN) Laboratory on the Washington State University Tri-Cities campus. His research concerns the use of emerging technologies in virtual reality, augmented reality, and neurocognition in decision-making, attitude formation, and problem-solving in students as well as simulations to foster learning. He also leads the development of a computer science education certificate program for K–12 teachers in Washington.

Snow Yunxue Fu
is an assistant arts professor at the Tisch School of the Arts in the Department of Photography and Imaging at New York University. She sees new media art making within a postphotographic framework as one of the most exciting interdisciplinary and relevant art fields of the 21st century. An internationally exhibited new media artist practicing for over a decade, Fu is active as a curator and collaborator with international artists and arts organizations focused on emerging simulation technologies, such as 3D, VR, AR, Web 3.0, and the metaverse, where her projects are hosted on social VR platforms like Sansar, Mozilla Hubs, and Decentraland.

David Glassmeyer
is a professor of mathematics education at Kennesaw State University, GA, USA, who teaches graduate mathematics and mathematics education courses in the online M.Ed., Ed.S., and Ed.D. programs in middle and secondary grades education. Dr. Glassmeyer aims to help teachers increase STEAM integration and consequently K–12 student learning. As a mathematics education researcher, he focuses on examining and developing teachers' STEM content knowledge, specifically secondary teachers' reasoning of mathematics concepts.

Luke Hampton
is an art educator and alumnus of the Visual Arts Administration program at New York University. He has over 15 years of experience in international art education and urban nonprofits, developing curricula for visual arts, writing, and technology in the United States and China and is active in the contemporary art world as a writer and collaborator with Snow Yunxue Fu. As an educator he currently trains international students to succeed as artists and professionals. As a researcher Hampton engages in the culture, histories, and current events that shape the experiences of artists and art students from China.

Kevin Hsieh
is a professor of art education at Georgia State University, GA, USA. Dr. Hsieh's research topics include instructional technology, museum education, LGBTQ+ issues, AsianCri pedagogy, and transdisciplinary art education. He has presented his research studies internationally. His articles have appeared in research journals, such as *Studies in Art Education, Art Education, Visual Arts Research, Museology Quarterly, Humanities and Social Sciences Research, Journal of Fine Arts, Journal of Liberal Arts, International Journal of Art and Design Education*, and *Social Sciences*.

Sean Justice
is an associate professor of art education at Texas State University in San Marcos, Texas. Dr. Justice teaches and writes about art and computer science education, creative computation, and teacher education in the digital age. As an artist, he has exhibited photographs, videos and computer animations both nationally and internationally. His book *Learning to Teach in the Digital Age: New Materialities and Maker Paradigms in Schools* was published by Peter Lang in 2016. In 2020 he received a National Science Foundation award to study teachers' computational thinking in early childhood classrooms with 4- to 8-year-old children.

Chia-Hung Kao
is an assistant professor from National Kaohsiung Normal University, specializing in art education and printmaking. Dedicated to promoting art education and practicing the spirit of A/r/tography, he was honored in 2021 with the Award for Excellence in Praxis by the International Society for Education through Art, a UNESCO affiliate. His printmaking works have garnered accolades, including the Gold Medal at the Taiwan Fine Arts Exhibition and the Kaohsiung Award. His works are collected by the Ministry of Culture and international galleries.

Alice Lai
is a professor at the School of Arts and Humanities at Empire State University, State University of New York, USA, where she formerly coordinated undergraduate online arts curriculum for 20 years. She earned a Ph.D. in art education from the Ohio State University and an M.A. in art education and a B.A. in graphic design from California State University at Los Angeles. She received the Susan H. Turben Award for Excellence in Scholarship at Empire State University and the Marantz Distinguished Alumni Lectureship Award from the Ohio State University. Her teaching topics include diversity in the arts, women's art history, art theory, and arts-based research. Her research topics include art education, STEAM education, Asian American transnational art, and critical theories and pedagogies encompassing feminism, critical multiculturalism, anticolonialism, and digital equality. She frequently publishes in the field of art education and delivers presentations at national and international conferences.

Christine Liao
(Ph.D.) is an associate professor and coordinator of the Art Education MAT program at the University of North Carolina Wilmington. Her research areas include not only theorizing virtual body and identity and exploring interactions of the virtual and the real but also media arts and digital performance, STEAM, technology integration in art education, and intercultural education. She has presented at national and international conferences, including NAEA, AERA, and InSEA, and has published in journals in the field of art education and education.

Don McMahon
is an associate professor of special education at Washington State University, who has proven experience leading research teams during the implementation of technology-based interventions for people with ID and ASD. Dr. McMahon

has conducted research on a variety of technology-based interventions for individuals with disabilities, including mobile devices, video modeling, communication tools, podcast-based read-aloud testing accommodations, augmented reality applications, and virtual reality. In summer 2011 he was selected to attend the Harvard Graduate School of Education summer institute on universal design for learning. Dr. McMahon has 12 years of experience involving inclusive postsecondary education. A cofounder of the WSU ROAR program, he has served as its director.

Isao Ohashi
is a professor emeritus of Okayama University as well as a professor of education at Wakayama Shinai University in Japan. His research focuses on the learning process in art as a progression toward self-actualization, the relationship between perception and representation, and teaching methods for education through art from infancy to adolescence. He has served as the representative director of the Art Education Society of Japan and president of the Japan Practical Art Education Society. His research has been published in his books on art education, as book chapters, and in numerous journals in the field of art education in Japan.

Vedat Özsoy
graduated from the Gazi Faculty of Education in Türkiye, completing his master's degree at Birmingham Institute of Art and Design in U.K., his Ph.D. at Gazi University, and his postdoctorate fellowship at Arizona State University in the USA. He achieved the rank of professor in 2004. Having founded the Visual Arts Education Association (GÖRSED) in 2003, he served as its president until 2010. Prof. Özsoy also served as the InSEA Secretary (2014–2017). In addition, he has published books and articles on art and design education, museum education, curriculum development, and art teacher education.

Eliza Pitri
earned a B.A. in early childhood education from the University of Cyprus and an M.A. and Ph.D. in art education from the University of Texas–Austin and The Ohio State University, respectively. She is currently an associate professor of art education in the Department of Education at the University of Nicosia, Cyprus, teaching in both the undergraduate and the graduate program. Her research interests focus on socioconstructivist learning through the visual arts in various contexts by facilitating meaningful artmaking related to processes and skills, such as constructing a knowledge base, playfulness, flexibility, risk-taking, fluency, originality, humor, problem finding, and problem solving.

Shyh-Shiun Shyu
is the chairperson of Hakka Affairs Commission, Taipei City Government, who obtained his master's degree in molecular genetics from the State University of New York at Buffalo. Before joining Taipei City Government, he was the Taipei City Councilor, focusing on digital education and antidrug programs for children. He is also a devoted, lifelong volunteer with autistic children and is now a board member of the Republic of China (ROC) Foundation for Autistic Children and Adults in Taiwan.

Debrah C. Sickler-Voigt
(Ph.D.) is a professor of art education at Middle Tennessee State University. She authored *Teaching and Learning in Art Education: Cultivating Students' Potential from Pre-K Through High School* (2020), a widely-adopted art methods textbook, followed by *STEAM Teaching and Learning Through the Arts and Design: A Practical Guide for PK–12 Educators* (2023), her second book. Sickler-Voigt served as the senior editor for NAEA's Assessment Papers for Art Education from 2016 to 2023. She received the Ziegfeld Service Award in 2022 from the United States Society for Education Through Art and the Southeastern Region Higher Education Art Educator Award in 2023 from the National Art Education Association.

Agni Stylianou-Georgiou
is an associate professor of educational psychology in the Department of Education at the University of Nicosia. She studied elementary education at the University of Cyprus and earned a Ph.D. in educational psychology (cognition and instruction) from the University of Connecticut. Her research interests focus on metacognition, and creative teaching and learning in face-to-face and digital environments. She has been involved in EU-funded projects targeting the role of information and communication technologies in enabling a pedagogy for autonomy for teacher professional development and fostering dialogue and argumentation for cultural literacy learning in schools using wordless picture books.

Michael White
is the interim director of the Ernest G. Welch School of Art & Design and an Associate Professor of Interior Design. He is both NCARB and NCIDQ certified and is a registered architect and interior designer in the State of Georgia with degrees from Mississippi State University and the Georgia Institute of Technology. Michael joined Georgia State following an extensive national career in interior architectural practice, most notably as Studio Director of the Atlanta office of Gensler—the nation's largest interior architecture firm. His

20+ years of professional experience in commercial interiors included more than one million square feet in projects for national clients including Aetna, the McDonald's Corporation, Bank of America, BlueCross BlueShield, GTE, and Atlanta's own legal powerhouse, King & Spalding.

So Lan Wong
is currently a senior lecturer and the program leader of the Professional Development Programme (Visual Arts Teaching) in the Department of Cultural and Creative Arts at the Education University of Hong Kong. Dr. Wong's research interests include visual arts curricula and instructional design, including STEAM teaching. She is currently a member of the author team of *Hong Kong Chronicles—Visual Arts* (《香港志：視覺藝術》; 2022–25), a consultant (Phases 1–3, 2020–2023), and the principal investigator (Phase 3) of the Special Education Needs Section of the Curriculum Development Institute of the Hong Kong Education Bureau. She is also a specialist in the Hong Kong Council for Accreditation of Academic and Vocational Qualifications (2018–2024) and the Vice Chairman of the World Chinese Art Education Association (2019–2025).

Yui-Chih Wu
is a Ph.D. student in art education at Taiwan Normal University and a licensed art teacher in Taiwan. She has worked at the Ju Ming Museum, where she established alliances between the museum and schools in New Taipei County. This experience led to her employment at Gymchina, an art education institution in China, where she sought to expand contemporary art into the realm of children. When she served as vice CEO of Taipei Hakka Cultural Foundation, she integrated contemporary art education into ethnic cultural education. Her work focuses on STEAM, arts-based research, and data visualization.

INTRODUCTION

Celebrating Diverse Perspectives on STEAM Education

Yichien Cooper and Alice Lai

In the evolving landscape of education, integrating science, technology, engineering, arts, and mathematics into a holistic and robust STEAM curriculum has garnered attention worldwide. Global citizens are now equipped with digital, technological, and scientific literacy coupled with divergent, flexible creative thinking (Gess, 2017). As the coeditors of the current volume, we (Yichien and Alice) have respectively embarked on journeys to prepare ourselves and our students for life in a global society advanced by STEAM knowledge and products. Since mid-2010, we have joined forces to conduct teaching, research, and writing projects to explore the concepts, practices, potential, and pitfalls of STEAM education (e.g., Lai & Cooper, 2016, 2023). This anthology is a result of our years-long efforts in gathering researchers, educators, policymakers, community organizers, and independent scholars in the United States and overseas to unveil diverse perspectives on STEAM concepts, teaching, and creativity.

Through our ongoing collaborative research, we have identified critical, social, and tactical aspects of STEAM learning. Examining gender divides emerging through and in visual culture, we uncovered young students' (ages 8–14) gender assumptions in which STEM tools and products were "boyish," or associated with boys, and items and practices involving arts and crafts were "girly," or associated with girls (Lai & Cooper, 2016). To transform such a mindset, we devised hands-on projects for students at a workshop, providing them with STEAM materials to create artworks, such as prototypes of a car and a porch, and fostering creativity, innovation, problem-solving aptitude, and sense of identity—STEAM approaches recognized by the National Art Education Association (2022). We extended our collaboration (Lai & Cooper, 2023), designing and implementing STEAM lessons through project-based learning (PBL). Our aim was to develop adolescent girls' STEAM thinking and the capabilities needed for civic engagement. We emphasized asking driving questions, meeting the standards in the disciplines involved, conducting authentic and professional inquiry, engaging in collaborative activities, and creating artifacts

 | DOI:10.1163/9789004714748_001

pointing toward possible answers to the driving question—PBL approaches now commonly applied in STEAM fields. As a result, the students used virtual-reality gadgets, a 3-D printer, and STEAM materials to create artifacts revealing their solutions to civic problems that require studies of animal science, energy conversation, urban aesthetics, and community life.

Living on opposite sides of the United States, we continue to dialogue about STEAM education. As art educators believing in the vital role of the "A" in STEAM education, we have strived to put forth conceptual and empirical studies of STEAM education in this volume along with ideas and field-tested samples to build the knowledge base undergirding STEAM pedagogy in our field. In the following sections Yichien shares insights into STEAM as a form of arts integration, and Alice outlines the trajectory and state of STEAM education in our field. Culminating with synopses of each chapter, we invite readers to go global and ponder the myriad facets of STEAM education.

1 STEAM as a Form of Arts Integration

In April 2013, the Next Generation Science Standards (NGSS) included Crosscutting Concepts (CCC), establishing deeper connections across seemingly separate disciplines in science and engineering practices. In the NGSS seven CCC s were identified to bridge core ideas and integrate common practices: (a) pattern; (b) cause and effect; (c) scale, proportion, and quantity; (d) systems and system models; (e) energy and matter (flow, cycles, and conservation); (f) structure and function; and (g) stability and change. In response, Rivet et al. (2016) applied Lakoff and Johnson's (2003) conceptual metaphors to linguistically represent this complex phenomenon. They categorized CCC s as lenses, bridges, tools, and rules, which conceptualize and highlight the functions of CCC s across teaching content, student activities, and learning outcomes.

As an instructor specializing in integrating the fine arts into curricula, Yichien was intrigued by this progressive approach that fosters holistic learning, aligning closely with arts disciplines and practices. Whether studio habits of mind (Hetland, 2007; Hogan et al., 2018)—namely, engage and persist, envision, express, reflect, question and explain, evaluate, sketch and explore, and understand art world—can be internalized in art classes and applied in other disciplines is worth pondering (Getting, 2016; Winner, 2022). Over time Yichien explored intersections of the arts and STEM, particularly through involvement in the maker movement, in which students engage in design thinking to bring their ideas to life. Her experience as a coach for the community-based FIRST® LEGO® League and Destination Imagination (DI) teams further deepened her understanding of how to cultivate students' creativity and potential in purposeful, collaborative, project-based learning environments.

On one hand, DI emphasizes the creative process and transformation of STEAM programs to enhance learners' problem-solving and self-expression skills. Through various Instant Challenges and cross-curricular activities, participants integrate multiple disciplines to generate ideas and meet challenges with limited resources. On the other hand, the FIRST® LEGO® League focuses on guiding young learners through STEM exploration, emphasizing discovery, innovation, impact, inclusion, teamwork, and fun as core values. Teams design and program robots to complete missions, followed by researching and developing prototypes to address real-world problems aligned with the mission theme. Yichien observed the presence of art and a studio mindset in both DI and FIRST® LEGO® League. The former actively promotes and celebrates the arts across curricula, while the latter, despite engaging students in engineering design processes and incorporating theatrical elements for presentations, does not explicitly mention the arts. Exploring the visible and hidden connections between art and STEM fuels her mission to integrate a comprehensive STEAM approach into education.

Considering STEAM a form of arts integration, Yichien expanded her previous work by constructing a learner-centered environment in which curricula are viewed as bridges connecting a learner's prior knowledge, social and cultural experiences, and new knowledge through exploration, interpretation, reflection, and creation (Cooper, 2013). She continues to explore the impact of STEAM in teacher training and learning—like many educators interested in this method—by experimenting with and blending various learning approaches. Early attempts at STEAM education are included in her works on integrating arts curricula for Chinese educators (陳怡倩, 2016, 2017a, 2017b, 2018).

Over more than a decade-long endeavor in STEAM research and teaching, Yichien has recognized that education rigor begins with creativity, which can be acquired through the following:

- Multimodality. Multiple modes of learning are used to engage students' sensory experience to convey learning concepts.
- Uniqueness. Learners' capabilities and strengths are recognized to connect learning.
- Interactivity. Knowledge relies on the interactivity of the brain. Cross-disciplinary integration helps spark ideas, sustain thinking, and invite connectivity in learning.

STEAM emphasizes the connection, interweaving, and integration of subjects and continuously cultivates learners' flexible thinking in various situations and contexts. To this end, the content of teaching requires strategies to stimulate students' learning potential. Creativity allows students to take part in learner-centered, purposeful exploration and discovery. For example, in Yichien's preservice teacher class, students discussed various ways to design a maze from LEGO blocks; then they compared the impact of teachers' guided approaches on the level of creative problem-solving. The first guiding approach involved students following the teacher's predesigned template with a clear step-by-step manual to complete the task; the second encouraged students to design their own maze, but it had to contain concepts of continuality, branches to direct or indirect paths, and obstacles; the third integrated gravity into the maze design concept mentioned in the second approach. Students designed a LEGO marble-run maze to explore conditioned movement, demonstrating their understanding of push and pull, the force of friction, and kinetic energy. Both the second and third approaches provided organic learning opportunities for creative problem solving; however, framed by a purposeful teaching goal, the third approach amplified students' ability to realize abstract learning concepts to apply interdisciplinary skills to address a real-life context (Drake & Burns, 2004; Sousa & Pilecki, 2018). In other words, different strategies implemented with the same teaching materials yielded polarized results in terms of the sustainability of learning.

To foster creative and inclusive learning environments, diverse STEAM curricula are needed. Three components for developing them include the following:

- Improvisational learning. Co-construct meaning through divergent thinking.
- Relational knowledge. Provide embodied experiences and make connections with self (brain, mind, and body) and others (social, cultural, and economic).
- Equitable access. Form respectful partnerships for culturally responsible teaching. Provide differentiated scaffolds in consideration of student needs and cultural heritage.

Diverse STEAM curricula allow flexibility and spontaneous happenings in learning, employing systematic and holistic approaches to transform object-centered and material-focused teaching methods into relationship-based, sense-making processes for all learners (Gaskins, 2021). The aspects of building relational knowledge in STEAM education mirrors the idea, people, object, and physical theory, or IPOP for short, found in museum settings, in which visitors' preferred museum experiences encompass object, cognitive, introspective, and social experiences (Perkarik et al., 2014).

Indeed, in a STEAM learning environment, individuals are naturally drawn to the intersections of various experiences across all dimensions. Project-based, participatory-centered, potential-driven, and process-focused STEAM curricula encourage learners to actively explore, interpret, reflect, and create meaning. Such curricula focusing on facilitating integrative, creative, and holistic experiences invite educators to rethink common STEAM practices that overly emphasize rigid machinery and technological skills constraining learners' creativity. Effectiveness in learning is not about saving time or providing a clear road map for students to follow passively. Instead, the emergence of STEAM education as an alternative learning approach is about building sustainable relationships among learners, concepts, and ability (Figure I.1).

Because of isolation and social distance regulations, the COVID-19 pandemic impacted STEAM education practice; however, educators became more resourceful and drew inspiration from many sources. During that time Yichien's family embarked on a STEAM exploration routine, enacting countless

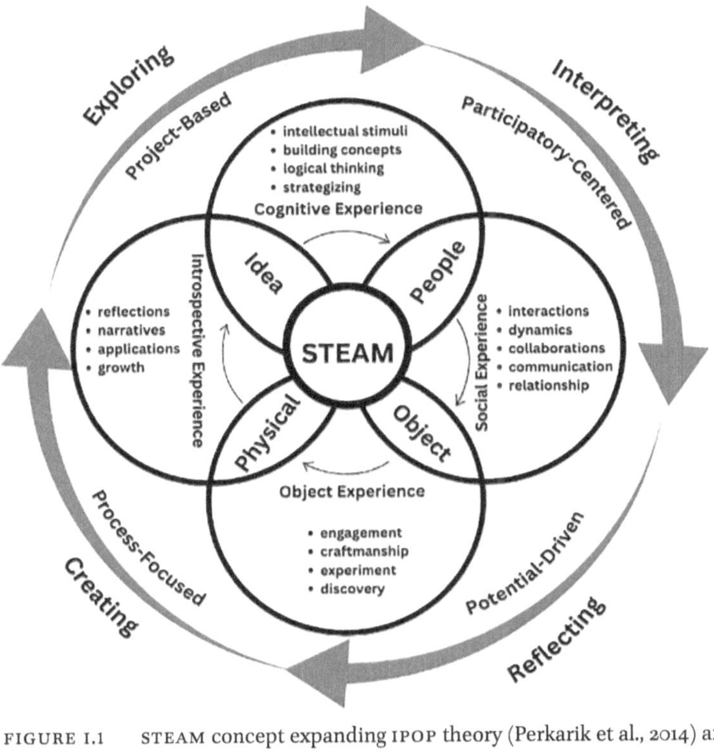

FIGURE I.1 STEAM concept expanding IPOP theory (Perkarik et al., 2014) and curricula as bridges (Cooper, 2013)

FIGURE 1.2 Pinhole photography, 2021. Exposure from January 10 to June 10, 2021

improvisations to accommodate each member's physical, cognitive, and mental readiness. Modifications allowed all members to research and discover relational knowledge across disciplines, prior experience, and social cultural and economic factors related to a teaching concept. For example, they studied a brief history of photography in addition to constructing pinhole cameras from soda cans to capture long exposure images in the backyard (Figure 1.2). They discussed darkroom experiences and the process of developing film prior to instant results from digitalized smartphones. Another unit engaged the children in extracting color from plants and discovering that irises, although providing pristine blue hues, produce strong, poisonous fumes. In addition to creating pixel art to reinforce concepts like coordinates, grids, and symmetry with LEGO or Minecraft, the children designed tensegrity sculptures to further their understanding of balance, tension integrity, and compression force (Figure 1.3). These appealing STEAM challenges employing the potential of each learner in design thinking and the creative problem-solving process also served as a place of confluence for idea brewing, materials tinkering, and bonding.

2 STEAM Education: A View from the Field of Art Education

STEAM approaches to knowledge construction and artmaking were recognized as long ago as the 14th-century Renaissance (Rolling, 2016); however, STEAM education as a pedagogical paradigm advocated through governmental efforts is still a burgeoning phenomenon in many countries. The trajectory of STEAM

FIGURE I.3 Example of tensegrity sculpture by Emilie Cooper, 2020

education in the United States was initially steered by a political and economic agenda to improve STEM education and to increase opportunities for competition and profit from STEM-driven international commerce. Since 2010, the increased call for the inclusion of the arts, encompassing various disciplines in the arts and humanities, into STEM education has led to the expansion of STEM to STEAM education. After a decade the STEAM approach was listed as one of the 10 common educational approaches emphasized by K–12 art educators (Bertling & Moore, 2021). As shown in Table I.1, a range of nationwide and discipline-based initiatives have contributed to the rapid growth and popularity of STEAM education in the field of art education (Gess, 2017).

The nationwide and discipline-based efforts to promote and improve STEAM education necessitated an internal survey to determine the response to these efforts in our field of art education. To this end, Alice conducted a literature review to gauge the research interests in STEAM education in the field. Her first task was to identify articles containing the keyword "STEAM" published between January 2010 and April 2024 in high-circulation and peer-reviewed journals in the field: *Art Education*, *Studies in Art Education*, *Visual Arts Research*, *International Journal of Education & the Arts*, and *International Journal of Art & Design Education*. Because the word STEAM has multiple meanings, Alice's second task was to apply three criteria to assess the level of relevance in regard to STEAM education in each article. Articles selected

TABLE I.1 Key initiatives advocating for STEAM education nationwide and in the field of art education

Year	Key initiative
2010	Former President Obama called a CEO-led effort to improve STEM education nationwide.
2012	John Maeda, the former president of the Rhode Island School of Design, spoke at the 2012 National Art Education Association (NAEA) Convention promoting the synergistic relationship between art and STEM and design as a STEM product.
2012	*Art Education* published the first article fully featuring the STEM–STEAM concept and activities by Bequette and Bequette (2012).
2013	Maeda continued lending his expertise, esteemed position and history in academia, and voice to various venues advocating for STEAM education, including a political move supporting the newly formed Congressional STEAM Caucus.
2014	NAEA established the first version of four position statements for STEAM education.
2015	Every Student Succeeds Act (2015) called for the inclusion of the arts in STEM Education.
2016	*Art Education* published two special issues on STEAM education.
2018	The theme of the NAEA convention was Art + Design = STEAM.

for further content analysis must feature (a) STEAM education as the central theme, (b) substantial discussion of the concept, history, policy, or literature review of STEAM education, and (c) pedagogical samples of or suggestions for future STEAM education. A STEAM-centered article meeting all three criteria was deemed "highly relevant." An article containing phrases or ideas related to STEAM education cursorily was deemed "relevant."

The preliminary data, as shown in Table I.2, revealed that among the journals surveyed, *Art Education* had the highest number (n=25) of highly relevant articles dedicated to diverse themes related to STEAM education. This journal primarily included empirical case studies and practical classroom ideas and samples shedding light on K–12 STEAM pedagogy, activities, and creativity. The second highest number (n=6) of highly relevant articles, albeit far below 25, appeared in the *International Journal of Education & the Arts*. Each of the

TABLE I.2 Preliminary data on STEAM-centered articles appearing in selected journals in the field of art education

Name of Journal (2010–2024)	# of STEAM-centered articles (highly relevant)	# of STEAM-centered articles (relevant)
Art Education	25	
International Journal of Education & the Arts	6	
International Journal of Art & Design Education	3	
Visual Arts Research		4
Studies in Art Education		2

rest of the journals had fewer than five highly relevant articles fully devoted to themes related to STEAM education.

Evidently, STEAM education has gained traction among K–12 art educators, yet how well STEAM education has corresponded to the expectations set forth by the field has remained a vital question to be addressed. To do so, Alice delved into a content analysis of the 34 STEAM-centered and highly reverent articles as they related to the four position statements for STEAM education newly revised by the NAEA (2022, para 3):

> Visual arts content and artistic ways of thinking are fundamental and valuable components of high-quality STEAM education. NAEA believes that
>
> 1. STEAM education values all STEAM disciplines equally.
> 2. STEAM education is implemented through a wide variety of approaches.
> 3. STEAM education encourages creativity and innovation and problem-solving.
> 4. STEAM education acknowledges the rigor found in the visual arts.

The results of the content analysis showed that Position Statements 2 and 3 were well-reflected in the STEAM-centered articles, featuring abundant discussions supporting the purpose and values of STEAM education to augment students' creativity, innovation, and problem-solving ability. Numerous

TABLE I.3 Various approaches applied in STEAM education

Art infused	Inquiry-based
Arts integrated	Problem or issue-based
Collaboration-based	Project-based
Community-based	Real world-based
Culturally relevant	Transdisciplinary
Design thinking	

pedagogical samples, instructional strategies, learning activities, and student artwork can be found in STEAM-centered articles, indicating ways that creativity, innovation, and problem-solving manifest in STEAM lessons as well as a wide array of recurring STEAM approaches employed in art, design, and art education classrooms (summarized in Table 1.3). Among them, transdisciplinarity, design thinking, and real world-based learning were some of the most frequently adopted pedagogical frameworks and approaches.

Position Statements 1 and 4 essentially emphasized the importance of teaching the content knowledge, discipline standards, and professional expectations of visual art in STEAM activities. The results of content analysis in regard to these two statements were mixed, indicating a need to improve the ways art is taught or incorporated in STEAM activities. On one hand, many STEAM-centered articles featured pedagogical samples and students' creative projects, affirming that art is a domain of knowledge taught in STEAM activities and that artistic learning can foster various types of thinking and academic skills required by STEM disciplines. On the other hand, a noticeable number of STEAM-centered articles revealed art as a stimulus for STEM learning, a motivator for girls to become interested in STEM subjects, a visual aid in STEM learning and projects, or a communication or presentation tool. In such examples, art was neither valued nor taught equally and rigorously among STEAM subjects.

A faculty member teaching primarily undergraduate arts and humanities courses, Alice surveyed research on STEAM in the field of art education, leading her to further reflect on her experience and the state of STEAM in higher education in the United States. Literature on STEAM teaching in higher education has not been booming compared to that in the K–12 context. The disparity has been attributed to the silo phenomenon in the modern higher education system that promotes the specialization of academic subjects and the compartmentalization of academic disciplines (Herro et al., 2019; Kraehe, 2018). Consequently, the primary responsibilities of university professors often include teaching classes in their own schools or departments, conducting research and publishing results in their own fields, and participating in professional activities benefiting their departments and fields. This methodically discipline-based ecosystem limits incentive or opportunity for professors to teach or engage in research and professional development activities beyond their fields. Some professors have, in fact, expressed feelings of inadequacy or frustration when teamworking with colleagues outside their fields, refused to collaborate with colleagues of dissimilar expertise, or resisted developing cross-disciplinary curricula altogether because of constraints in time, funding, and professional obligation (Herro et al., 2019). Siloed U.S. higher education foments a phenomenon in which each academic department offers a curriculum mostly responding to its discipline requirements, and professors concentratedly emphasize knowledge and skills conforming to trends in their disciplines.

In her own experience as a graphic design and art education student and an arts and humanities professor, Alice noticed that undergraduate and graduate programs seldom feature a cross-disciplinary curriculum that integrates various disciplines beyond the arts and humanities. Although students may take some science, technology literacy, and mathematics classes to fulfill general education requirements for their degrees, such subjects are taught in a compartmentalized mode, hence rarely connecting to art or stimulating students' thinking about art. In certain studio art classes—photography, figure drawing, painting, ceramics, sculpture, and graphic design—knowledge of chemistry, anatomy, the vitrification process, engineering, mathematical perspectives, geometry, algebra, calculus, and physics are essential in ensuring the completion of durable and visually effective art works of quality. Professors, however, tend to simplify or omit the scientific and mathematical explanations in these studio art classes, resulting in students creating artworks following the prescribed instructions and formulas or mimicking existing work without grasping the basic STEM knowledge that makes their creations possible and aesthetically intriguing. In art education, appreciation, and theory classes, the concept of art as a cross-disciplinary or transdisciplinary construct intersecting with STEM, although not new, has not been consistently emphasized; consequently, students are deprived of opportunities to develop a complex and inclusive view of art and to examine a wide array of STEAM artists and artworks.

The literature review and observation of STEAM education in the K–16 context indicate the following shortcomings and needs that future researchers on STEAM education may want to consider:

- Research on STEAM education in undergraduate arts and humanities classes has been sporadic. Students of art education develop foundational

art knowledge and skills in such classes; hence, more research in this area is necessary.
- More studies on the challenges STEAM educators encounter, especially in compartmentalized educational settings, are needed to assist K–16 educators and administrators in overcoming institutional barriers that hinder cross-departmental and cross-disciplinary STEAM collaboration.
- According to the NAEA "effective STEAM education embraces the National Visual Arts Standards and standards from other content areas" (2022, para 1). More research showing inclusion and analysis of standards from visual art as well as involved STEM areas is needed.
- Additional studies on the assessment of STEAM learning and projects are needed, but because STEAM education is not a single discipline-based learning paradigm, researchers and policymakers may need to ponder new standards and assessment schemes reflecting integrative STEAM learning necessary to assess it genuinely and holistically.
- STEAM education is seen as disciplinary transcendent, or transdisciplinary; that is, STEAM learning should generate new knowledge and product beyond the sum of the involved disciplines. Researchers should indicate what new knowledge and product, especially with respect to art, is generated as the outcomes of STEAM learning.
- Whether or not extended STEAM education has elevated students' so-called 21st-century skills and enhanced their learning or making of art remains ambiguous, theoretical, or anecdotal. More comprehensive or longitudinal research on this is needed.

3 STEAM Education: A Global Phenomenon

With an extensive scholarly record on STEAM education, since 2016 Yichien has had the privilege of conducting curriculum design and STEAM workshops for educators and museum staff across China, Malaysia, Taiwan, Türkiye, Canada, and the United States. She has visited myriad classrooms, observing educators from various disciplines embracing the concept of STEAM education. Notably, Ban Chiao High School in Taipei, Taiwan, has dedicated a building to incorporate career and technical education, the FIRST Robotics Competition, and a maker space (Make4fun) for hands-on, project-based design challenges, partnering with the National Palace Museum for creative cultural connections. Many Taiwanese media arts educators have emphasized the use of technology to facilitate the creative process in virtual reality, film, filmmaking, and graphic design (Cooper, 2022).

In her travels Yichien has seen that STEAM education gained ground globally through institutional and governmental initiatives. In Penang, Malaysia, Chung Ling Private High School has integrated STEAM through machinery and coding. In Türkiye, her meeting with then President of the Scientific and Technological Research Council of Turkey, or TÜBİTAK, Professor Hasan Mandal prompted dialogue with prominent STEM educators and researchers, advocating for design thinking in the arts to enhance traditional STEM education. In Thailand, the practice of art was extended to the fields of medicine and healthcare, with students using artistic and technological skills for the enhancement of visual communication. Meanwhile, in Shanghai, Beijing, and Hong Kong, Chinese art educators embraced STEAM to foster creativity and innovation, moving away from conventional learning methods, reflecting their focus on sustainable economic development and international compatibility. STEAM education has become a global phenomenon. Leading the effort in producing this anthology with Alice, Yichien has continued to be on the lookout for the development of STEAM education in the United States and overseas.

4 The Aims and Structure of STEAM Education: Intersections and Thresholds

The aim of this book was to broaden the scope of STEAM education by exploring its profound meaning, emphasizing sustainable teaching and learning experiences. We use the term "intersections" to invite collaborative possibilities where ideas from multiple disciplines converge. In addition, we highlight "thresholds" as a metaphor for embarking on new beginnings characterized by insightful educational practices and innovative happenings.

Given the rapid advancements in science and technology, educators are prompted to rethink knowledge transfer and application, especially in the current discourse on artificial intelligence. The field of education is at a critical juncture, when the need for innovative and flexible teaching approaches surpasses traditional modes of learning that catered to passing standardized tests; therefore, gathering insights and experiences from STEAM practitioners globally to navigate these challenges is crucial. This book includes authors from diverse educational backgrounds contributing unique and varied perspectives and expertise to the implementation of STEAM education. From elementary school teachers and community arts advocates to university professors from Asia, Europe, and the United States, each author enriches the STEAM paradigm by showcasing distinctive interpretations and applications.

This book comprises four sections.

1. *Rethinking STEAM Frameworks.* This section underscores the essence of STEAM education, providing diverse philosophical and educational perspectives. Christine Liao (Chapter 1) advocates for teaching STEAM through transdisciplinary competencies, emphasizing empathy in the design process to promote collaboration across disciplines. Embracing these strategies, Liao aims to cultivate critical thinkers and conscientious global citizens among students. Debrah C. Sickler-Voigt (Chapter 2) introduces an authentic STEAM teaching model through art and design, emphasizing artistic behaviors, interdisciplinary studies, teacher–student choice, and hands-on learning for idea development, observation, imagination, persistence, and creative connections. In exploring the intersection of art education and STEAM in Japan, Isao Ohashi's (Chapter 3) analysis underscores the vital role of art in fostering creativity, innovation, intrinsic value, and self-actualization within STEAM education. It highlights educational practices such as *zoukei asobi* (artistic play), expressive behavior, cultivating a creative attitude, and promoting self-actualization. Vedat Özsoy (Chapter 4) examines the efforts and challenges of transitioning from STEM to STEAM education in Türkiye, focusing on promising initiatives by the Ministry of National Education as well as key limitations in Türkiye's STEAM education developments.
2. *Redirecting STEAM Foci in the Communities.* This section features a range of alternative STEAM practices in communities, bringing critical issues like sustainability, responsible global citizenship, cultural competencies, and accessibility to light. Paul Boey (Chapter 5) discusses the state of STEAM education in Penang State, Malaysia, emphasizing the value of both school-based and community-based STEAM projects despite challenges. Chia-Hung Kao (Chapter 6) illustrates how he developed three STEAM modules to foster interdisciplinary learning of natural science, art, and aspects of Taiwanese culture, emphasizing collaborative STEAM skills acquisition and enjoyment among students. Eliza Pitri, Maria Evagorou, and Agni Stylianou-Georgiou (Chapter 7) reflect on the development and refinement of a STEAM curriculum for 12- to 14-year-old girls by practicing four pedagogical principles: learning with purpose, empowerment and agency, immersive learning, and interdisciplinary blending through design thinking. Shyh-Shiun Shyu and Yui-Chih Wu (Chapter 8) share their implementation of a STEAM-infused curriculum, where children explored the cultural, historical, and family data of the Hakka ethnic group in Taiwan through data visualization. This approach fosters both family-centered emotional exploration and ethnic collective

memory while enhancing data literacy skills. Hsin-Yi Chao (Chapter 9) examines the future directions of visual arts education for people with visual impairments and explores ways to foster a social environment conducive to inclusive learning through STEAM applications.

3. *Reimagining STEM as STEAM.* This section features the intersections of art and STEM and an array of best practices in transforming STEM into STEAM education. Jonah B. Firestone and Don McMahon (Chapter 10) showcase augmented reality as it enables students and teachers to overlay 3D objects, text, and animations onto the real world, allowing for interactive manipulation using smartphones. David Glassmeyer and Kevin Hsieh (Chapter 11) outline the integration of art into mathematics education to illustrate the pendulum dynamics and the distinctions between prime and composite numbers. Snow Yunxue Fu and Luke Hampton (Chapter 12) focus on Fu's use of virtual reality, Web 3.0 technologies, and collaborative learning strategies to prepare students for future creative and technological advancements. Sean Justice and Lori Czop Assaf (Chapter 13) explore the intersections of art, literacy, and creative computing within the framework of STEAM, emphasizing how creative computing enhances story creation and expressive meaning making in collective interactions with texts. Alice Lai (Chapter 14) introduces the arts-driven STEAM (ADSTEAM) model to foster students' transdisciplinary learning of aesthetic theories and making of art. Drawn from historical and Bauhaus perspectives, a studio activity prompting students to apply Fibonacci mathematical patterns to create visual work illustrates the potential of ADSTEAM learning.
4. *Remapping Teacher Education Through STEAM Practices.* This section centers on ways to reinstall creativity for competency building in teacher preparation programs to resist overemphasis on standardized tests. Suzan Duygu Bedir Erişti and Yeliz Erdoğan (Chapter 15) detail their process of planning STEAM-oriented instructional activities in visual arts education to train future teachers in content knowledge across disciplines. So Lan Wong (Chapter 16) delineates STEAM activities for preservice teachers, demonstrating ways the arts were integrated into STEM learning to foster creativity and knowledge in the involved disciplines leading to a critical review of STEAM education in Hong Kong. This section concludes with Kevin Hsieh and Michael White's (Chapter 17) experience of including skill-based and concept-based components to train STEM teachers, highlighting the significance of positive teacher attitudes toward art integration for schools promoting STEAM curricula.

5 STEAM Education for the Future of Humanity

While we developed this anthology, COVID-19 ravaged the globe, resulting in escalated human dependence on the use of technology with blooming online learning platforms. Education that emphasizes technology and neglects arts and humanities, however, will never achieve the goal of whole-child education. We believe the rise of the STEAM educational paradigm has created an opportunity for educators to rethink the idea of the rigor of 21st-century education and to prepare students for their future in a global society. The emergence of STEAM education is thus both a reminder and an expectation, prompting us to uncover and share the significances and inner workings of STEAM education.

At its core, STEAM education represents a departure from traditional siloed learning, fostering interdisciplinary connections and promoting holistic understanding. By blending scientific inquiry with artistic expression, technological innovation with engineering design, and mathematical reasoning with creative problem-solving, STEAM learning transcends the boundaries of individual subjects to create a more integrated and immersive learning experience. By merging these traditionally distinct disciplines, proponents of STEAM education seek to nurture well-rounded individuals equipped with the critical skills and competencies needed to thrive in an increasingly interconnected and rapidly evolving world.

Attested by authors in this anthology, STEAM education represents a transformative approach to expand pedagogical paradigms and learning experiences. The value of arts education must surpass positively impacting academic performances and extend to cultivating personal growth in competency building, well-being with empathy toward one another, and socioeconomic prospects (Venkatesh et al., 2023). As we move forward to provide quality education for the next generation, we hope this volume offers gateways and opportunities to envision STEAM education in diverse learning environments.

References

Bequette, J. W., & Bequette, M. B. (2012). A place for art and design education in the STEM conversation. *Art Education*, *65*(2), 40–47. https://doi.org/10.1080/00043125.2012.11519167

Bertling, J. G., & Moore, T. C. (2021). The U.S. K–12 art education curricular landscape: A nationwide survey. *Studies in Art Education*, *62*(1), 23–46. https://doi.org/10.1080/00393541.2020.1858007

Cooper, Y. (2013). Curriculum connections: Mapping integrating arts framework for diverse learnings. *International Journal of Arts Education, 11*(1), 184–199.

陳怡倩. (2016). 課程設計: 統整藝術課程設計的思維與趨勢 [*On integrated curriculum design*]. Hungyeh Publishing Co.

陳怡倩. (2017a). 統整的力量 [*The power of integrated curriculum: The core of STEAM education*]. Hunan Publishing Co.

陈怡倩. (2017b). 从STEAM的A来看美国的STEAM教育 [The role of A in STEAM education],香港美术教育, *1*, 4–9.

陈怡倩. (2018). 透析STEAM教育的核心理论. [Theories behind STEAM education]. 中国中小学美术, *131*(July), 52–55.

Cooper, Y. (2022). The use of time in media arts education: Stories from Taiwan. In A. Knochel & S. Osamu (Eds.), *Global media arts: Mapping perspectives of media arts in education*. (pp. 307–322). Palgrave Macmillan.

Drake, S., & Burns, R. (2004). *Meeting standards through integration curriculum*. Association for Supervision and Curriculum Development.

Every Student Succeeds Act, 20 U.S.C. § 6301. (2015). https://www.congress.gov/bill/114th-congress/senate-bill/1177

Gaskins, N. R. (2021). *Techno-vernacular creativity and innovation: Culturally relevant making inside and outside of the classroom*. MIT Press.

Gess, A. H. (2017). STEAM education: Separating fact from fiction. *Technology and Engineering Teacher, 77*(3), 39–41.

Gettings, M. (2016). Putting it all together: STEAM, PBL, scientific method, and the studio habits of mind. *Art Education, 69*(4), 10–11. https://doi.org/10.1080/00043125.2016.1176472

Herro, D., Quigley, C., & Cian, H. (2019). The challenges of STEAM instruction: Lessons from the field. *Action in Teacher Education, 41*(2), 172–190. https://doi.org/10.1080/01626620.2018.1551159

Hetland, L. (2007). *Studio thinking: The real benefits of visual arts education*. Teachers College Press.

Hogan, J., Hetland, L., Jaquith, D., & Winner, E. (2018). *Studio thinking from the start: The K–8 art educator's handbook*. Teachers College Press.

Kraehe, A. M. (2018). Disciplinary borderlands. *Art Education, 71*(2), 4–7. https://doi.org/10.1080/00043125.2018.1414528

Lai, A., & Cooper, Y. (2016). Untangling gender divides through girly and gendered visual culture. *Journal of Social Theory in Art Education, 36*, 96–107.

Lai, A., & Cooper, Y. (2023). We are VR girls: Becoming engaged citizens through STEAM-integrated projects. In M. Bae-Dimitriadis & O. Ivashkevich (Eds.), *Teaching civic participation with digital media in art education: Critical approaches for classrooms and communities* (pp. 151–164). Routledge. https://doi.org/10.4324/9781003402060-14

Lakoff, G., & Johnson, M. (2003). *Metaphors we live by* (2nd ed.). University of Chicago Press.

National Art Education Association. (2022, June 1). *NAEA position statement on STEAM education.* https://www.arteducators.org/advocacy-policy/articles/552-naea-position-statement-on-steam-education

Next Generation Science Standards. (2013, April 10). *Appendix G – Crosscutting concepts.* https://www.nextgenscience.org/sites/default/files/Appendix%20G%20-%20Crosscutting%20Concepts%20FINAL%20edited%204.10.13.pdf

Pekarik, A. J., Schreiber, J. B., Hanemann, N., Richmond, K., & Mogel, B. (2014). IPOP: A theory of experience preference. *Curator*, *57*(1), 5–27. https://doi.org/10.1111/cura.12048

Rivet, A. E., Weiser, G., Lyu, X., Li, Y., & Rojas-Perilla, D. (2016). What are Crosscutting Concepts in science? Four metaphorical perspectives. In C. K. Looi, J. L. Polman, U. Cress, and Reimann, P. (Eds.), *Transforming learning, empowering learners: The International Conference of the Learning Sciences (ICLS)*, V. 2, (pp. 970–973). International Society of the Learning Sciences.

Rolling, J. H. (2016). Reinventing the STEAM engine for art + design education. *Art Education*, *69*(4), 4–7. https://doi.org/10.1080/00043125.2016.1176848

Sousa, D. A., & Pilecki, T. (2018). *From STEM to STEAM: Brain-compatible strategies and lessons that integrate the arts* (2nd ed.). Corwin. https://doi.org/10.4135/9781544357393

Venkatesh, V., Ruprecht, L., & Ferede, M. K. (2023). *Arts education: An investment in quality learning*. United Nations Educational, Scientific and Cultural Organization. https://unesdoc.unesco.org/ark:/48223/pf0000386034

Winner, E. (2022). *An uneasy guest in the schoolhouse: Art education from colonial times to a promising future*. Oxford University Press. https://doi.org/10.1093/oso/9780190061289.001.0001

PART 1

Rethinking STEAM *Frameworks*

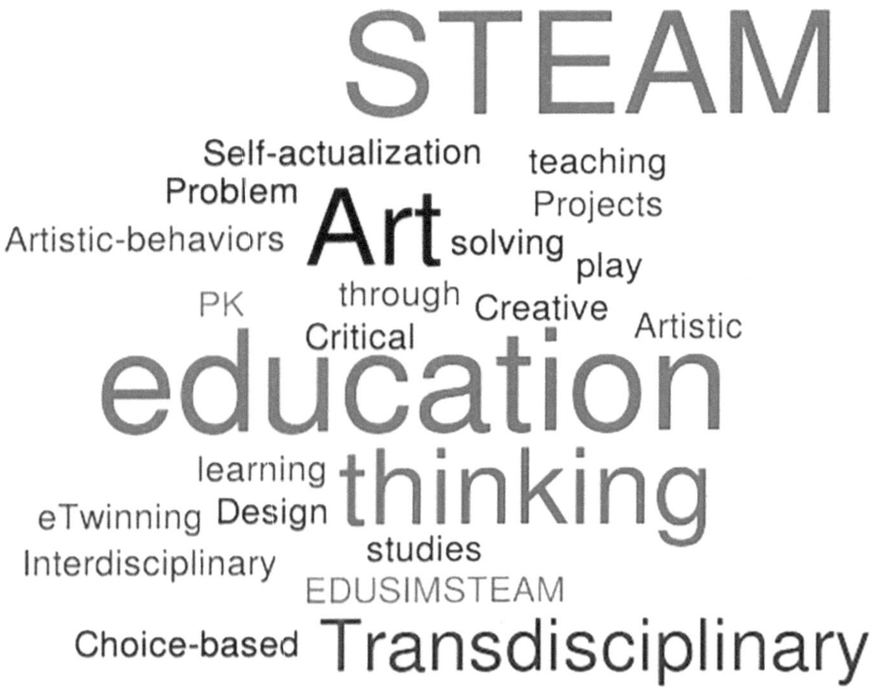
STEAM
Self-actualization
teaching
Problem
Art
Projects
Artistic-behaviors
solving
play
PK
through
Creative
Artistic
Critical
education
learning
thinking
eTwinning
Design
studies
Interdisciplinary
EDUSIMSTEAM
Choice-based
Transdisciplinary

CHAPTER 1

You Don't Need a List of the Best STEAM Lesson Ideas

Focusing on the Transdisciplinary Competencies

Christine Liao

Abstract

In this chapter I recommend teaching STEAM through transdisciplinary competencies, including critical thinking, problem solving, creative thinking, and social and emotional skills. I offer several suggestions for effectively teaching these competencies, which include incorporating opportunities to foster critical thinking in the curriculum, nurturing essential attitudes and dispositions for both critical thinking and social and emotional learning, employing design thinking to teach problem-solving skills, prioritizing empathy as an integral part of the design thinking process, and facilitating collaboration within and across disciplines. By teaching transdisciplinary competencies, students can develop into critical thinkers and responsible global citizens.

Keywords

transdisciplinary – critical thinking – creative thinking – problem solving – design thinking

1 Introduction

Teachers wishing to teach science, technology, engineering, arts, and mathematics (STEAM) may start by searching for ideas online. A quick Google search can provide hundreds of thousands of STEAM lesson ideas, all claiming to be the most fun and engaging. But do these "best lesson ideas" truly teach the critical STEAM skills that students need to know?

Teaching STEAM is not about cherry-picking individual lessons or ideas from the Internet or other resources. It is about building a transdisciplinary learning environment in which all students can learn and explore STEAM

 | DOI:10.1163/9789004714748_002

concepts and build transferable knowledge and skills (Bertrand & Namukasa, 2020; Suganda et al., 2021). This cannot be accomplished by merely following the lesson plans of others; rather, it requires carefully thought out pedagogy that involves diverse students. In this chapter I recommend, teaching STEAM through transdisciplinary competencies, including but not limited to critical thinking, problem solving, creative thinking, and social and emotional skills; I also offer suggestions for designing STEAM curricula to teach these competencies.

2 The Current Problem in STEAM Education

STEAM education has seen a rise in popularity in the last several years and has been adopted in classrooms around the United States and many other countries (Chu, 2021; Deák & Kumar, 2022). The hope is that by integrating these areas of study, students will be better equipped for the future and develop their ability to think critically and creatively (Rahmawati et al., 2019; Suganda et al., 2021; Wilson et al., 2021). As a result of the increasing popularity of STEAM education, many new lesson plans, tools, materials, and curricula have been developed for K–12 schools; however, several issues have emerged in the current STEAM education landscape.

Because no universally accepted definition of STEAM education is in place, the quality of STEAM lessons and curricula can vary considerably; anyone can claim that a particular curriculum or program falls under the STEAM umbrella based on personal interpretations. Integration is seen as crucial to STEAM education; however, many teachers simply "throw in" art or STEM components without truly integrating them (Liao, 2016). Many "STEAM lessons" available on the Internet appear to be based on preexisting lessons with added art or STEM components. For example, a lesson with a cutout of a skeleton template for coloring, supposed to teach science and art, is sold as a "STEAM craft activity" on the popular teacher resources site *Teachers Pay Teachers* (Skeleton Puppet—Multiple Use—STEAM Craft Activity, n.d.). This example shows that some educators may use STEAM as a buzzword without changing their teaching methods or curricula. In addition, many educators and schools see STEAM education or art as an "add-on" or "extracurricular" rather than a core part of the educational experience (*New Study Gives Guidance*, n.d.). These educators are merely following a new trend, not creating a meaningful integration of STEAM subjects.

In addition to these issues, misconceptions about STEAM education abound, particularly with the integration of art into STEM. Some educators are convinced that the STEM disciplines will be weakened by the addition of art (Gallagher & Rawson, 2019). Others think that STEAM education simplifies STEM courses and adds the arts to attract more women into STEM fields (Kijima & Sun, 2021); furthermore, the arts are erroneously perceived as focusing only on the final product rather than the process (Perignat & Katz-Buonincontro, 2019). Finally, many teachers think that "STEAM teaching is a set of interesting add-on activities" (Jamil et al., 2018, p. 415).

Despite the problems and concerns, studies have shown the advantages of STEAM education in a variety of ways, particularly when STEAM education is approached from the viewpoint of transdisciplinary learning—blurring the boundaries between disciplines (e.g., Guyotte et al., 2015; Liao, 2016; Wilson et al., 2021). As the number of educators focusing on teaching Social and Emotional Learning (SEL) increases, it is important not to overlook STEAM in this trend. Although STEM subjects are not usually associated with soft skills, such as relationship skills emphasized in SEL, many similar skills, including collaboration and responsible decision-making, are taught in both STEAM and SEL (Lammert & Drummond, 2021).

STEAM education can improve student engagement and academic accomplishment even though instructors must make substantial changes in their teaching methods to successfully incorporate and develop STEAM instruction (Quigley & Herro, 2016). Notwithstanding the challenges associated with implementing STEAM classrooms, it is a vital and beneficial educational method that will continue to be a part of the education landscape; therefore, rather than asking where to acquire resources for those "amazing" STEAM lessons on the Internet, teachers should ask what competencies and abilities they want to teach via STEAM education. By viewing STEAM education through a transdisciplinary lens, teachers can create meaningful and engaging lessons that truly teach students important skills and concepts.

3 Defining STEAM as a Transdisciplinary Educational Approach

To return to one of the primary reasons for these problems and misconceptions—the lack of a clear definition of STEAM education—many scholars have undertaken analyses to identify and categorize the various STEAM education practices in an effort to find a more precise interpretation of STEAM and illuminate the best path forward (e.g., Liao, 2019; Mejias et al., 2021; Perignat & Katz-Buonincontro, 2019; Quigley et al., 2017). For example, through an analysis of current practices, Perignat and Katz-Buonincontro (2019) identified five ways to differentiate STEAM education: transdisciplinary, interdisciplinary, multidisciplinary, cross-disciplinary, and arts integration. Among them, they recommended a transdisciplinary approach. Mejias et al. (2021) advocated for transdisciplinary STEAM practice rooted in "social purpose and meaning-making" (p. 225)—meaningfully benefiting the community and society—to transcend the transdisciplinary approach. In sum, most studies have shown benefits, such as developing transferable skills—critical thinking and problem solving, promoting cooperative learning, and engaging in self-discipline—associated with a transdisciplinary STEAM approach in education (e.g., Clark & Button, 2011; Costantino, 2018; Liao, 2016).

Transdisciplinarity differs from multidisciplinarity or interdisciplinarity. Multidisciplinarity refers to the practice of merging or engaging with more than one academic discipline; the focus of interdisciplinarity is on the links that may be made between various fields of study, but transdisciplinarity opens the door to new possibilities because it is "at once between the disciplines, across the different disciplines, and beyond all discipline [*sic*]" (Nicolescu, 1997, para. 4). In other words, it involves pushing beyond the traditional boundaries of disciplines and developing new fields of research, education, application, and much more outside those boundaries. It is not merely about crossing disciplines but about a process of creating new knowledge that cannot be found in any one discipline. It is also an approach to thinking, a way of critically examining the assumptions behind the categories used to categorize the world. Transdisciplinarity, therefore, requires critical thinking, creativity and creative thinking, problem solving, curiosity, and flexibility, to name a few (Mokiy, 2019; Wagner et al., 2014).

4 Transdisciplinary Competencies in STEAM

Many educational reforms have focused on teaching competencies to help students acquire future-relevant skills and knowledge (Evans & Vander Els, 2018; Moon, 2007). Consequently, many educators have shifted from focusing on transmitting knowledge to emphasizing the development of competencies. Several studies have proven that STEAM can teach students various skills crucial to them in today's rapidly changing environment, such as critical thinking, problem solving, creativity and creative thinking, and collaboration skills (Dyer, 2019; Wagner et al., 2014). These skills are commonly referred to as "21st-century skills" (Trilling & Fadel, 2009); however, many use the terms without clearly defining what they are. In fact, a variety of definitions for these skills are held among scholars (Ramalingam et al., 2020). Holbrook et al. (2020)

attempted to identify transdisciplinary skills and group them into five categories: communication, research, thinking, social, and self-management. Despite the lack of consensus on the transdisciplinary competencies STEAM teaches, in what follows, I highlight three of the most widely discussed skills—critical thinking, creative thinking, and problem solving—and their connections to STEAM education.

4.1 *Critical Thinking*

Critical thinking can be generally defined as a cognitive process involving various abilities to analyze and evaluate information to arrive at a conclusion and make a judgment (Ennis, 2015). Critical thinking skills typically involve "analyzing arguments, making inferences using inductive or deductive reasoning, judging or evaluating, and making decisions or solving problems" (Lai, 2011, pp. 41–42). In addition to these skills, critical thinking also involves dispositions—attitudes or habits of mind—such as "open- and fair-mindedness, inquisitiveness, flexibility, a propensity to seek reason, a desire to be well-informed, and a respect for and willingness to entertain diverse viewpoints" (Lai, 2011, p. 42). Critical thinking is essential for making good decisions and discovering answers to problems and challenges, and it can be taught through art classes. For example, Lampert (2006) stated that in art classes, teachers encourage critical thinking in their students by encouraging them to examine several solutions to creative challenges and by having them reflect on the unique viewpoints of their peers through classroom critiques.

Studies have demonstrated that STEAM education improves students' critical thinking skills (Weng et al., 2022; Wilson et al., 2021). Critical thinking is best taught by allowing students to practice in real-world situations. Project-based transdisciplinary STEAM education teaches critical thinking skills by encouraging students to ask questions, discover issues, and find solutions in a real-world context (Maeda, 2012; McGarry, 2018). Thus, students can better build their own problem-solving strategies, assess information, and make judgments based on their analyses (Hadinugrahaningsih et al., 2017). STEAM education provides students with opportunities to exercise critical thinking skills to evaluate the consequences of their choices so they can understand that their choices affect both the environment and others (Bassachs et al., 2020).

4.2 *Creative Thinking*

Creative thinking can be characterized as "the capacity to generate many different kinds of ideas, manipulate ideas in unusual ways, and make unconventional connections in order to outline novel possibilities that have the potential to elegantly meet a given purpose" (Ramalingam et al., 2020, p. 2). Simply put, creative thinking is the capacity to generate new ideas, innovations, or fresh perspectives. Art educators have long shared ways to teach creative thinking through art education (Marshall, 2010), for example, encouraging "fluency, flexibility, originality, and elaboration" (Heid, 2008, p. 46) and promoting design thinking (Wiebe & Seaton, 2021).

One of the arguments often cited for combining STEM and the arts to form STEAM is that art fosters creative thinking (Land, 2013; Maeda, 2013). Transdisciplinary STEAM education has been shown to cultivate creative problem solving, according to several studies (Guyotte et al., 2015; Suganda et al., 2021; Wilson et al., 2021). STEAM education encourages students to think creatively by providing opportunities to apply what they have learned in novel contexts (Sangngam, 2021). Creative thinking encourages students to think and solve problems from multiple perspectives; as a result, they can become better problem solvers and innovators.

4.3 *Problem Solving*

Although the term "problem-solving skill" may be understood in a variety of ways depending on the field, in general it refers to an individual's capacity to recognize a problem, uncover its root cause, consider a variety of potential solutions, and put those answers into action (Martinez, 1998). Problem-solving skills are critical for success in any field. Understanding the context that contributes to the problem and comprehending the issue are two important aspects of problem solving. Sometimes solving a problem calls for creative or abstract thinking (Dumper et al., 2019).

The problem-solving skills essential to success in the 21st century are best taught through a transdisciplinary STEAM approach, where students not only learn about a range of fields and how they relate to one another but also how knowledge and ideas can be generated in the transdisciplinary space (Kim & Lee, 2022). STEAM education provides opportunities for students to learn to solve problems, especially when students are engaged in the design thinking process (Henriksen, 2017; Watson, 2015). Design thinking is perceived as a problem-solving method that focuses on developing students' creative confidence (Rauth et al., 2010; Ulibarri et al., 2014). The students participate in projects designed to help them develop empathy, encourage creativity, promote an attitude toward action, and stimulate the development of active problem-solving skills (Rusmann & Ejsing-Duun, 2022). In this process students learn to take a problem, identify its key components, and develop solutions. Despite some variety in terms used in the design thinking process, the process generally includes six stages: empathize (understand people's needs and motivations), define (understand the problem to identify key issues), ideate (generate innovative ideas), prototype (make a rough version of the final product), test (test prototypes in real life to gather feedback), and implement (bring the design concept to life; Gibbons, 2016). Because design thinking may help students think critically and address real-world challenges and because students benefit from the materiality and authenticity of it (Rusmann & Ejsing-Duun, 2022), using it in STEAM education is a popular approach to ensuring students gain problem-solving skills and experiences that equip them for the real world (Gross & Gross, 2016; McGarry, 2018; Watson, 2015).

5 Connecting STEAM Learning with Competencies in Social and Emotional Learning

Social and emotional learning (SEL) addresses five core competencies: self-awareness, self-management, social awareness, relationship skills, and responsible decision-making (*What Is the CASEL Framework?*, n.d.). The connection between art and social and emotional learning is well-recognized; for example, using art to express emotions is one way teachers can engage students in conversations about emotions. Studies have shown that art can teach social and emotional learning by helping young people develop empathy, self-awareness, and other social and emotional learning skills (De, 2018; Hanning, 2022). A connection between social-emotional learning and STEAM also exists, primarily through mutually constructed shared competencies, such as social and interpersonal skills and cognitive regulation (e.g., critical thinking; Bennett, 2021). Students who receive a STEAM education have been shown to have higher social and emotional learning skills levels than those who do not (Peterson et al., 2018). On the contrary, researchers have predicted that students with highly developed social and emotional learning skills are also likely to be more successful in STEAM fields in the future (Guo et al., 2022). This can be understood in terms of the transferability of these competencies. STEAM education connects to social and emotional learning in a variety of ways. In Figure 1.1, I illustrate the complex relationships among STEAM teaching approaches (design thinking, collaboration, and project-based learning), STEAM competencies, and SEL skills represented as concentric circles. The primary STEAM competencies (critical thinking, creative thinking, and problem solving) are all part of the competencies taught through these approaches. Each approach is also linked to a closely related SEL skill (self-awareness, self-management, social awareness, relationship skills, and responsible decision-making). The competencies and approaches are layered, signifying their

FIGURE 1.1 The relationship between STEAM teaching approaches, STEAM competencies, and SEL skills

interconnectedness and overlap. In what follows, I describe the connections separately in the five SEL skills, but awareness of their interconnection is important.

5.1 *Self-Awareness and Self-Management*

A well-designed STEAM curriculum allows students to explore and discover knowledge in areas that interest them without compelling them to follow a curriculum that does not meet their needs or interests. This kind of curriculum allows students to pursue their own interests and passions, which may lead to a greater knowledge of themselves. It also helps to foster self-awareness as students become increasingly aware of their skills, interests, and capabilities (Heid, 2008; Marshall, 2010).

Students may acquire skills linked to self-management via STEAM education by establishing goals for themselves and accomplishing those objectives through hard work and dedication (Holbrook et al., 2020). As a result, they are able to learn how to manage their behavior, emotions, and impulses. Education in the STEAM disciplines also teaches students how to concentrate on a single activity or project for a long time. For example, students need to focus their attention in order to solve math problems or spend a long time working on an art project. This kind of experience is important because it helps students learn to be patient and manage their attention. This relationship among STEAM learning, self-awareness, and self-management is represented in the center of Figure 1.1.

5.2 *Social Awareness and Relationship Skills*

One of the effective approaches to teaching STEAM is through design thinking to engage students in a problem-solving process. Empathy—understanding and experiencing others' perspectives—is an essential competency and step in design thinking. Designers must empathize with users and understand their needs and perspectives to create meaningful designs (Lee & Park, 2021). Students can learn to stand in their peers' shoes by designing and creating a solution to a problem at their school (Rusmann & Ejsing-Duun, 2022). By partaking in STEAM education through the design thinking process, students will acquire more social awareness, improve their abilities to collaborate with others, and learn new approaches to finding answers to challenges. Problem-solving skills are essential to social and emotional development because students need to solve problems not just in academic tasks but in self-management and in relationships with others as well as in a collaborative environment (Totan & Kabasakal, 2012). STEAM teaches students how to collaborate to find solutions to issues and finish tasks, which in turn increases their capacity to cooperate with other people.

Many STEAM curricula are designed to help students develop their creativity and critical thinking skills by creating projects that can be used in the real world (Peng & Kueh, 2022). This encourages students to make a connection between the classroom and real life, which helps them develop their capacity for empathy. By considering the potential effects of their projects on the real world, students become more aware of how others might react or respond in similar situations and gain a deeper awareness of others. This perspective nurtures essential social awareness and relationship-building skills. In addition, incorporating real-world challenges into STEAM curricula can help students recognize the importance of collaboration and effective communication, which are integral to both social–emotional learning and the STEAM fields.

5.3 *Responsible Decision-Making*

Several researchers have indicated that STEAM enables students to practice decision-making skills (Ge et al., 2015; Holbrook et al., 2020), which require critical thinking and problem-solving skills. As noted in the previous section, through hands-on project-based learning students can learn to recognize the consequences of their decisions and how to evaluate them properly. For example, at the elementary school level, a project asking students to design and operate a school-wide recycling program to address environmental concerns and promote social responsibility and decision-making can be a good way to teach STEAM content and decision-making. Throughout the project students need to make decisions, such as selecting locations and designing recycling areas, that could impact their daily activities at school. This type of real-world project enables students to observe how their decisions can influence their interactions with others and the environment. By directly experiencing the result of their decisions, students can learn to make better decisions, also emphasized in social and emotional learning.

6 Recommendation: Focusing on Transdisciplinary Competencies

I have discussed the transdisciplinary competencies, including critical thinking, creative thinking, and problem-solving skills, that are fostered via STEAM education. The following are five recommendations for educators considering implementing STEAM education in their classrooms.

First, I recommend that teachers check their curriculum to ensure they include opportunities for fostering critical thinking skills, such as if their curriculum includes processes for students to analyze a variety of arguments, evaluate information, and solve problems (Lai, 2011). Instead of handing students the content knowledge, providing information from a range of sources to facilitate their critical thinking skills and evaluate the information to form conclusions is a step forward to include critical thinking in the curriculum. A STEAM curriculum can provide opportunities to engage students in these processes by adopting a project-based learning approach so students can analyze the facts and information and apply their knowledge to make choices and take action in an authentic setting.

Second, because fostering dispositions is also essential for critical thinking, a STEAM curriculum should contain components that nurture mindsets and attitudes that promote critical thinking, such as open-mindedness and fair-mindedness, curiosity, flexibility, respect, and a willingness to consider diverse perspectives (Lai, 2011). These attitudes have a connection with social and emotional learning as well. For instance, respect and willingness to consider other perspectives are part of social awareness skills in SEL (*T-SEL Competencies*, 2021). Teachers can encourage students to respect others and their viewpoints by deliberately emphasizing these qualities. For example, group work provides an excellent opportunity to practice respecting others' opinions. Bringing students with diverse backgrounds and knowledge together can teach them to learn others' perspectives in many contexts.

Third, design thinking should be used to teach problem-solving skills. To integrate design thinking within the existing curriculum, teachers can examine how each step of the process can be taught in a relevant context. For example, they can include brainstorming sessions, mind mapping, or group discussions in the lesson to guide students in analyzing information and synthesizing ideas to develop a clear problem statement to teach them to define problems and develop critical thinking. Offering an opportunity for prototyping will allow students to create a model of the designs that helps them identify opportunities for improvement and solve problems.

Fourth, prioritizing empathy as a component of the design thinking process is necessary to teach students social awareness as a part of SEL. The ability to understand and share other people's feelings is an essential skill for students to learn because it allows individuals to see beyond their own perspective and acknowledge others' needs. This capacity also plays a key role in conflict resolution because it helps people understand one another's perspectives before making decisions about how to move forward together. Students may learn to understand the emotions of others via the empathy process in design thinking, which involves having them design solutions to problems by putting themselves in the place of others who experience those problems. A context-rich lesson that engages students in solving real-world challenges or finding solutions for real people can cultivate this ability (Paracha et al., 2019).

Fifth, I recommend facilitating transdisciplinary competencies by encouraging collaboration within and across disciplines as much as possible. Collaboration is a skill important in both SEL and STEAM. As a shared competency, teaching SEL skills, such as self-management, relationship skills, and social awareness, can also help students build transdisciplinary collaboration ability, such as communicating across disciplines, understanding the perspectives of others, and being open to diverse ideas and viewpoints. Teachers can design collaboration projects that require students to draw upon their knowledge and skills from various disciplines to solve complex, real-world problems. In addition, creating partnerships with external organizations, such as community centers, local businesses, and nonprofit organizations, can further enhance the collaborative and transdisciplinary aspects of the STEAM curriculum. By engaging in projects that involve these external partners, students can gain valuable insights into the real-world applications of their learning, develop their collaboration skills, and establish connections with their community. Students will acquire an understanding of transdisciplinarity and how subjects are not separated in the real world.

When educators teach STEAM through transdisciplinary competencies, they not only help students develop essential skills for critical thinking and problem-solving but also foster necessary social and emotional skills because STEAM and SEL share common competencies. Through collaboration with peers on transdisciplinary projects, students learn to work together effectively. By empathizing with others to design solutions, they develop social awareness and concern for others. All of these skills are important not only in academic settings but also in life outside school. It is through teaching these transdisciplinary competencies, such as critical thinking, creative thinking, and problem solving, that educators help students become critical thinkers and responsible global citizens.

References

Bassachs, M., Cañabate, D., Nogué, L., Serra, T., Bubnys, R., & Colomer, J. (2020). Fostering critical reflection in primary education through STEAM approaches. *Education Sciences, 10*(12), 384. https://doi.org/10.3390/educsci10120384

Bennett, A. E. (2021). *Teachers' conceptualizations and experiences with STEAM instruction in the early childhood classroom in connection with children's social emotional competencies* (Publication No. 28547879) [Doctoral dissertation, Clemson University]. ProQuest Dissertations and Theses Global.

Bertrand, M. G., & Namukasa, I. K. (2020). STEAM education: Student learning and transferable skills. *Journal of Research in Innovative Teaching & Learning, 13*(1), 43–56. https://doi.org/10.1108/JRIT-01-2020-0003

Chu, H.-E. (2021). Editorial: STEAM Education in the Asia Pacific region. *Asia-Pacific Science Education, 7*(1), 1–5. https://doi.org/10.1163/23641177-bja10026

Clark, B., & Button, C. (2011). Sustainability transdisciplinary education model: Interface of arts, science, and community (STEM). *International Journal of Sustainability in Higher Education, 12*(1), 41–54.

Costantino, T. (2018). STEAM by another name: Transdisciplinary practice in art and design education. *Arts Education Policy Review, 119*(2), 100–106. https://doi.org/10.1080/10632913.2017.1292973

De, M. (2018). *Visual art's contribution to development of social and emotional skills and competencies: Impact analysis of an art-based social and emotional learning curriculum* (Publication No. 10931829) [Doctoral dissertation, Indiana University]. ProQuest Dissertations and Theses Global.

Deák, C., & Kumar, B. (2022, June 8). *Competence building in STE(A)M learning, innovation and entrepreneurship with students perspective on art and science in Europe* [Conference session]. The XXXIII ISPIM Innovation Conference "Innovating in a Digital World," Copenhagen, Denmark.

Dumper, K., Jenkins, W., Lacombe, A., Lovett, M., & Perimutter, M. (2019). *Introductory psychology*. https://opentext.wsu.edu/psych105/

Dyer, M. (2019). STEAM without hot air: Strategy for educating creative engineers. *Australasian Journal of Engineering Education, 24*(2), 74–85. https://doi.org/10.1080/22054952.2019.1693122

Ennis, R. H. (2015). Critical thinking: A streamlined conception. In M. Davies & R. Barnett (Eds.), *The Palgrave handbook of critical thinking in higher education* (pp. 31–47). Palgrave Macmillan US. https://doi.org/10.1057/9781137378057_2

Evans, C. M., & Vander Els, J. (2018). *Competency-based education: Educational reform. A primer* (Policy Brief #18-01; pp. 1–12). Education Department, University of New Hampshire. https://www.researchgate.net/publication/324780787_Competency-Based_Education_Educational_Reform_A_Primer

Gallagher, C. M., & Rawson, C. H. (2019). Art. In C. H. Rawson (Ed.), *Picking up STEAM: Science, technology, engineering, arts, and mathematics instruction in the public library*. https://steam.web.unc.edu/chapters/chapter-6-art/

Ge, X., Ifenthaler, D., & Spector, J. M. (2015). Moving forward with STEAM education research. In X. Ge, D. Ifenthaler, & J. M. Spector (Eds.), *Emerging technologies for STEAM education* (pp. 383–395). Springer International Publishing. https://doi.org/10.1007/978-3-319-02573-5_20

Gibbons, S. (2016, July 31). *Design Thinking 101*. Nielsen Norman Group. https://www.nngroup.com/articles/design-thinking/

Gross, K., & Gross, S. (2016). Transformation: Constructivism, design thinking, and elementary STEAM. *Art Education, 69*(6), 36–43.

Guo, J., Tang, X., Marsh, H. W., Parker, P., Basarkod, G., Sahdra, B., Ranta, M., & Salmela-Aro, K. (2022). The roles of social–emotional skills in students' academic and life success: A multi-informant and multicohort perspective. *Journal of Personality and Social Psychology, Online*. https://doi.org/10.1037/pspp0000426

Guyotte, K. W., Sochacka, N. W., Costantino, T. E., Kellam, N. N., & Walther, J. (2015). Collaborative creativity in STEAM: Narratives of art education students' experiences in transdisciplinary spaces. *International Journal of Education and the Arts, 16*(15), 1–38.

Hadinugrahaningsih, T., Rahmawati, Y., & Ridwan, A. (2017). Developing 21st century skills in chemistry classrooms: Opportunities and challenges of STEAM integration. *The 4th International Conference on Research, Implementation, and Education of*

Mathematics and Science: Research and Education for Developing Scientific Attitude in Sciences and Mathematics, 1868(1). https://doi.org/10.1063/1.4995107
Hanning, K. (2022). Protection projects: How art on a cart reframed pedagogical practice to support social–emotional growth. *Art Education, 75*(2), 19–26. https://doi.org/10.1080/00043125.2021.2009285
Heid, K. (2008). Creativity and imagination: Tools for teaching artistic inquiry. *Art Education, 61*(4), 40–46. https://doi.org/10.1080/00043125.2008.11652067
Henriksen, D. (2017). Creating STEAM with design thinking: Beyond STEM and arts integration. *The STEAM Journal, 3*(1), 1–11. https://doi.org/10.5642/steam.20170301.11
Holbrook, J., Rannikmäe, M., & Soobard, R. (2020). STEAM education—A transdisciplinary teaching and learning approach. In B. Akpan & T. J. Kennedy (Eds.), *Science education in theory and practice* (pp. 465–477). Springer International Publishing. https://doi.org/10.1007/978-3-030-43620-9_31
Jamil, F. M., Linder, S. M., & Stegelin, D. A. (2018). Early childhood teacher beliefs about STEAM education after a professional development conference. *Early Childhood Education Journal, 46*(4), 409–417. https://doi.org/10.1007/s10643-017-0875-5
Kijima, R., & Sun, K. L. (2021). 'Females don't need to be reluctant': Employing design thinking to harness creative confidence and interest in STEAM. *International Journal of Art & Design Education, 40*(1), 66–81. https://doi.org/10.1111/jade.12307
Kim, S.-W., & Lee, Y. (2022). Developing students' attitudes toward convergence and creative problem solving through multidisciplinary education in Korea. *Sustainability, 14*(16), Article 9929. https://doi.org/10.3390/su14169929
Lai, E. R. (2011). *Critical thinking: A literature review research report.* Pearson. http://images.pearsonassessments.com/images/tmrs/CriticalThinkingReview FINAL.pdf
Lammert, C., & Drummond, S. (2021). Social and emotional teaching for STEAM through read-aloud: Kindergarteners learning to "feel courage-able." In J. Tussey & L. Haas (Eds.), *Advances in educational technologies and instructional design* (pp. 238–259). IGI Global. https://doi.org/10.4018/978-1-7998-7464-5.ch012
Lampert, N. (2006). Enhancing critical thinking with aesthetic, critical, and creative inquiry. *Art Education, 59*(5), 46–50. https://doi.org/10.1080/00043125.2005.11651611
Land, M. H. (2013). Full STEAM ahead: The benefits of integrating the arts Into STEM. *Procedia Computer Science, 20*, 547–552. https://doi.org/10.1016/j.procs.2013.09.317
Lee, H., & Park, J. E. (2021). Designing a new empathy-oriented prototyping toolkit for the design thinking process: Creativity and design sensibility. *International Journal of Art & Design Education, 40*(2), 324–341. https://doi.org/10.1111/jade.12345
Liao, C. (2016). From interdisciplinary to transdisciplinary: An arts-integrated approach to STEAM education. *Art Education, 69*(6), 44–49. https://doi.org/10.1080/00043125.2016.1224873

Liao, C. (2019). Creating a STEAM map: A content analysis of visual art practices in STEAM education. In M. S. Khine & S. Areepattamannil (Eds.), *STEAM education: Theory and practice* (pp. 37–55). Springer Nature.

Maeda, J. (2012). *STEM to STEAM: Art in K–12 is key to building a strong economy.* Edutopia. http://www.edutopia.org/blog/stem-to-steam-strengthens-economy-john-maeda

Maeda, J. (2013). STEM + Art = STEAM. *The STEAM Journal, 1*(1), Article 34. https://doi.org/10.5642/steam.201301.34

Marshall, J. (2010). Thinking outside and on the box: Creativity and inquiry in art practice. *Art Education, 63*(2), 16–23. http://dx.doi.org/10.1080/00043125.2010.11519057

Martinez, M. E. (1998). What is problem solving? *The Phi Delta Kappan, 79*(8), 605–609.

McGarry, K. (2018). Making partnerships with STEAM. *Art Education, 71*(2), 28–34. http://dx.doi.org/10.1080/00043125.2018.1414535

Mejias, S., Thompson, N., Sedas, R. M., Rosin, M., Soep, E., Peppler, K., Roche, J., Wong, J., Hurley, M., Bell, P., & Bevan, B. (2021). The trouble with STEAM and why we use it anyway. *Science Education, 105*(2), 209–231. https://doi.org/10.1002/sce.21605

Mokiy, V. S. (2019). International standard of transdisciplinary education and transdisciplinary competence. *Informing Science: The International Journal of an Emerging Transdiscipline, 22*, 73–90. https://doi.org/10.28945/4480

Moon, Y. (2007). Education reform and competency-based education. *Asia Pacific Education Review, 8*(2), 337–341. https://doi.org/10.1007/BF03029267

New study gives guidance on STEAM education integration. (n.d.). University of Pittsburgh School of Education. https://www.education.pitt.edu/new-study-gives-guidance-on-steam-education-integration/#:~:text=Key%20Findings%3A,innovation%20in%20real%2Dworld%20applications

Nicolescu, B. (1997, November). *The transdisciplinary evolution of the university: Condition for sustainable development.* CIRET: Centre International de Recherches et Études Transdisciplinaires. http://ciret-transdisciplinarity.org/bulletin/b12c8.php

Paracha, S., Hall, L., Clawson, K., Mitsche, N., & Jamil, F. (2019). Co-design with children: Using participatory design for design thinking and social and emotional learning. *Open Education Studies, 1*(1), 267–280. https://doi.org/10.1515/edu-2019-0021

Peng, F., & Kueh, C. (2022). Integration of design thinking with cultural intelligence in higher education for a socially complex environment. *International Journal of Art & Design Education, 41*(2), 341–354. https://doi.org/10.1111/jade.12402

Perignat, E., & Katz-Buonincontro, J. (2019). STEAM in practice and research: An integrative literature review. *Thinking Skills and Creativity, 31*, 31–43. https://doi.org/10.1016/j.tsc.2018.10.002

Peterson, A., Gaskill, M., & Cordova, J. (2018). Connecting STEM curriculum with social emotional learning (SEL) in elementary education. In E. Langran & J. Borup (Eds.),

Proceedings of Society for Information Technology & Teacher Education International Conference (pp. 1212–1219). Association for the Advancement of Computing in Education. https://www.learntechlib.org/primary/p/182681/

Quigley, C. F., & Herro, D. (2016). "Finding the joy in the unknown": Implementation of STEAM teaching practices in middle school science and math classrooms. *Journal of Science Education and Technology*, *25*(3), 410–426. https://doi.org/10.1007/s10956-016-9602-z

Quigley, C. F., Herro, D., & Jamil, F. M. (2017). Developing a conceptual model of STEAM teaching practices. *School Science and Mathematics*, *117*(1–2), 1–12. https://doi.org/10.1111/ssm.12201

Rahmawati, Y., Ridwan, A., Hadinugrahaningsih, T., & Soeprijanto. (2019). Developing critical and creative thinking skills through STEAM integration in chemistry learning. *Journal of Physics: Conference Series*, *1156*. https://doi.org/10.1088/1742-6596/1156/1/012033

Ramalingam, D., Aderson, P., Duckworth, D., Scoular, C., & Heard, J. (2020). *Creative thinking: Definition and structure*. Australian Council for Educational Research. https://research.acer.edu.au/ar_misc/43

Rauth, I., Köppen, E., Jobst, B., & Meinel, C. (2010). Design thinking: An educational model towards creative confidence. *Proceedings of the 1st International Conference on Design Creativity*, 1–8. https://hpi.de/fileadmin/user_upload/fachgebiete/meinel/papers/Design_Thinking/2010_Rauth_ICDC.pdf

Rusmann, A., & Ejsing-Duun, S. (2022). When design thinking goes to school: A literature review of design competences for the K-12 level. *International Journal of Technology and Design Education*, *32*(4), 2063–2091. https://doi.org/10.1007/s10798-021-09692-4

Sangngam, S. (2021). The development of early childhood students' creative thinking problem solving abilities through STEM Education learning activities. *Journal of Physics: Conference Series*, *1835*(1), 012008. https://doi.org/10.1088/1742-6596/1835/1/012008

Skeleton Puppet—Multiple Use—STEAM Craft Activity (Lesson Plan & Printable). (n.d.). *Teachers Pay Teachers*. https://www.teacherspayteachers.com/Product/Skeleton-Puppet-Multiple-Use-STEAM-Craft-Activity-Lesson-Plan-Printable-4897403

Suganda, E., Latifah, S., Irwandani, Sari, P. M., Rahmayanti, H., Ichsan, I. Z., & Rahman, M. M. (2021). STEAM and Environment on students' creative-thinking skills: A meta-analysis study. *Journal of Physics: Conference Series*, *1796*(1). https://doi.org/10.1088/1742-6596/1796/1/012101

Totan, T., & Kabasakal, Z. (2012). The effect of problem solving skills training on the social and emotional learning needs and abilities of 6th grade students. *Elementary Education Online*, *11*(3), 813–828.

Trilling, B., & Fadel, C. (2009). *21st century skills: Learning for life in our times.* Jossey-Bass.

T-SEL competencies: Social awareness. (2021, November 24). California Department of Education. https://www.cde.ca.gov/ci/se/tselsocialawareness.asp

Ulibarri, N., E. Cravens, A., Cornelius, M., Royalty, A., & Svetina Nabergoj, A. (2014). Research as design: Developing creative confidence in doctoral students through design thinking. *International Journal of Doctoral Studies, 9*, 249–270. https://doi.org/10.28945/2062

Wagner, T., Baum, L., & Newbill, P. (2014). From rhetoric to real world: Fostering higher order thinking through transdisciplinary collaboration. *Innovations in Education and Teaching International, 51*(6), 664–673. https://doi.org/10.1080/14703297.2013.796726

Watson, A. D. (2015). Design thinking for life. *Art Education, 68*(3), 12–18.

Weng, X., Cui, Z., Ng, O.-L., Jong, M. S. Y., & Chiu, T. K. F. (2022). Characterizing students' 4C skills development during problem-based digital making. *Journal of Science Education and Technology, 31*(3), 372–385. https://doi.org/10.1007/s10956-022-09961-4

What Is the CASEL Framework? (n.d.). CASEL. https://casel.org/fundamentals-of-sel/what-is-the-casel-framework/

Wiebe, M., & Seaton, S. (2021). Art education and design thinking: Reflections on students engaging with community using design. *The Canadian Review of Art Education, 48*(1), 3–19. https://doi.org/10.26443/crae.v48i1.98

Wilson, H. E., Song, H., Johnson, J., Presley, L., & Olson, K. (2021). Effects of transdisciplinary STEAM lessons on student critical and creative thinking. *The Journal of Educational Research, 114*(5), 445–457. https://doi.org/10.1080/00220671.2021.1975090

CHAPTER 2

Cultivating Authentic Approaches to STEAM Teaching and Learning through the Arts and Design

Debrah C. Sickler-Voigt

Abstract

This chapter presents a model of authentic approaches to cultivate STEAM teaching and learning through the arts and design. Its main components include (a) artistic behaviors, (b) interdisciplinary and transdisciplinary studies, (c) teacher and student choice, and (d) hands-on learning. Specifically, the artistic behaviors emphasize five habits: idea development, observation, imagination and wonderment, persistence, and making creative connections. Drawing examples from my book *STEAM Teaching and Learning Through the Arts and Design: A Practical Guide for PK–12 Educators*, I explain how this model of authentic approaches can benefit interdisciplinary and transdisciplinary STEAM studies in which the STEAM disciplines inform one another and can holistically expand beyond integrated subjects to form new understandings. Using historic and contemporary international artists' and designers' teaching and creative work as inspiration, I address the value of designing a choice-based STEAM curriculum and articulate the importance of hands-on learning in STEAM education.

Keywords

PK–12 STEAM education – interdisciplinary and transdisciplinary studies – artistic behaviors – choice-based teaching and learning

1 Introduction

STEM and STEAM subject matter abounds in PK–12 education (ages 4–18). Since 2010 the Rhode Island School of Design (n.d.) in the United States has championed the integration of the arts into STEM (science, technology, engineering, and mathematics) programs to provide students with a STEAM education in which the necessity of the arts and design is recognized in STEM innovations. Without the arts STEM neither presents a complete picture of what students

 | DOI:10.1163/9789004714748_003

need to know nor can the four areas fully meet the needs of society (Mejias et al., 2021; Sickler-Voigt, 2023). The innovative design solutions, aesthetic value, and creativity that the visual arts and design cultivate through STEAM bring STEM to life. National policies addressing a STEM/STEAM education have been implemented by the U.S. government. In 2015, the U.S. Congress reauthorized Elementary and Secondary Education Act, the United States' public education law, including its Every Student Succeeds Act (2015) that articulates the role of the arts in education. This act supports hands-on learning and explains how the arts benefit STEM education by "integrating other academic subjects, including the arts, into STEM subject programs to increase participation in STEM subjects, improve attainment of skills related to STEM subjects, and promote well-rounded education" (Every Student Succeeds Act, 2015, p. 1978).

Given the broad and open-ended educational policies towards STEM/STEAM, educators have formed different understandings of what a STEAM education can be (MacDonald et al., 2020; Mejias et al., 2021; Perales & Aróstegui, 2021). Mejias et al.'s (2021) research explained how STEAM's varying interpretations are due to its role as a nascent educational field, and so "it is often deployed in theory, pedagogy, and practice in ambiguous or potentially problematic ways toward varying ends" (p. 209). Thus, the authentic integration of the arts into STEM has sometimes been misinterpreted (Mejias et al., 2021; Sickler-Voigt, 2023). For example, as I conducted research on STEAM education to determine which resources were readily available for educators, I accessed numerous online STEAM lesson plans, as they are a common go-to source for educators (Topalska, 2021). I discovered that many of these lesson plans incorporated art materials and leisure crafting techniques. Their foci appear to be on generating student products that are fun to create but replicate same results. Consequently, their implementation minimizes the depth of student learning in the arts and design.

Some STEAM lessons apply forced integration. They call upon arts specialists to teach stereotypical arts content with the intent of only serving STEM disciplines. For example, in a science-focused lesson on papermaking, students identify the scientific qualities of paper and learn how to produce it; but its historical development, cultural applications, and associated vocabulary are neglected. Such lessons are often void of high-quality visual exemplars and fail to explain the relationships between the disciplinary content of STEM and the intrinsic value of the arts, their meaning, and rich history. Some educators misinterpret teaching the STEAM curriculum as possible only through the use of expensive technologies and kits. Further concerns occur when educators aiming to foster students' comprehensive understanding of the visual arts feel ill-equipped to teach STEAM due to lack of access to training in which the arts are treated as a group of disciplines coequal with STEM subjects (Allina, 2018; Perales & Aróstegui, 2021; Topalska, 2021). In fostering a comprehensive STEAM education, Mejias et al. (2021) have advocated for educators and policy makers "to combine arts and the STEM fields as valued and engaged coequally, to advance learning within, across, and beyond field boundaries" (p. 223).

In my professional role as an art education professor specializing in teacher education, curriculum design, and assessment, I recognize the challenges teachers face with STEAM education and have embarked on designing an authentic model anchored in best practice to support educators who are unfamiliar with STEAM education and/or who have struggled with teaching the arts in context. At the same time, I desire to provide holistic resources to experienced practitioners craving additional in-depth STEAM content to suit their program's needs and their own professional development interests. These efforts resulted in a resource book I wrote, *STEAM Teaching and Learning Through the Arts and Design*, for all educators teaching in STEAM disciplines and for the purpose of shaping educational policy on STEAM instruction (Sickler-Voigt, 2023).

Authentic approaches to STEAM education are fundamental because they serve as global passports that propel the curriculum to boundless destinations and advance a wide array of skills, mindsets, and behaviors that have lasting values in students' lives (Dubek et al., 2021; ElSayary, 2021; MacDonald et al., 2020). I argued that through "authentic" teaching, learning, and assessment practices, teachers and students identify real-world problems and solutions. Students learn worthwhile life skills that hold intrinsic and extrinsic value and motivate them by bringing joy, creativity, and humanist values to learning tasks (Perales & Aróstegui, 2021; Topalska, 2021); furthermore, they learn about the inspiration that drives artists, designers, and other STEAM professionals. Curricular tasks incorporate big ideas—"broad topics that students study in depth to address significant human issues that remain relevant regardless of the times and cultures in which people live" (Sickler-Voigt, 2020, pp. 6–7). In what follows, drawing ideas and examples from my STEAM book, I explain a model that facilitates these holistic practices using four components: (a) artistic behaviors, (b) interdisciplinary and transdisciplinary studies, (c) teacher and student choice, and (d) hands-on learning (Sickler-Voigt, 2023).

1.1 *Cultivating Artistic Behaviors*

Artists and designers seek to make sense of the world through artistic behaviors as they inform their practice and are driven by their training, experience, and intuition (Hetland et al., 2013; Sickler-Voigt, 2020). Like fellow STEAM professionals, artists and designers ask questions, form hypotheses, make predictions, analyze data, and generate informed interpretations to communicate ideas, foster understanding, and create design solutions. I have specifically focused on five artistic behaviors and the ways they can be taught and applied comprehensively in the classroom: (a) idea development, (b) observation, (c) imagination and wonderment, (d) persistence, and (e) making creative connections. The following sections describe these behaviors and include snapshots of some of the artists featured in my STEAM book, including Alma Thomas, Katsushika Hokusai, George Lukas, Litha Soyizwapi, Sally Blakemore, Antoni Gaudi, and Violet Oakley. These exemplars assist teachers and students in fusing STEAM subject matter with artists' and designers' thematic works and artistic behaviors and help them develop their own unique approaches to the STEAM curricular studies.

1.1.1 Idea Development

Idea development refers to the formation of artists' creative, innovative, and original ideas. Artists and designers build upon their prior knowledge and intuition. They employ reasoning skills to craft fresh ideas and apply metacognition—thinking about their own thinking processes—to make self-reflections and self-assessments about the quality of their work (Krathwohl, 2002). To facilitate their generation of ideas and research practices, they develop brainstorming lists, sketches, and models.

A comprehensive STEAM curriculum utilizes academic standards and builds enduring understanding to foster the development of students' ideas (Dubek et al., 2021). Art education scholar Eliot Eisner (1998) explained: "Human knowledge is a constructed form of experience and therefore a reflection of mind as well as nature: Knowledge is made, not simply discovered" (p. 7). Teaching artist Alma Thomas (1891–1978) demonstrated this concept. Throughout her life, she pursued her acquisition of knowledge by surrounding herself with art professionals. She refined her artistic skills by taking classes, participating in professional development, and practicing techniques at home (Smithsonian Institution Archives of American Art, 2019). Deeply interested in technology, Thomas painted her *Space Series* (1970–1974) to illustrate her ideas and interpretations of the scientific phenomena and achievements of her time as abstract art forms made with bright colors and striped patterns called Alma stripes.

1.1.2 Observation

The term "observation" describes artists' and designers' abilities to look closely, notice similarities and differences, search for clues, and discover new information. Eisner (1998) explained: "In the visual arts, to know depends upon the ability to see, not merely to look" (p. 6). Using the skill of close observation, artists and designers analyze and interpret what they see and transform their observational studies into creative works.

Students' observational skills can be developed through participating in art inquiry tasks, making life drawings, taking detailed notes in a journal, and playing memory games. Discovering visual data through the active process of locating or collecting original and reproduced artworks, designs, visual culture exemplars, models, sketches, and associated photographs via "scavenger hunts" on the internet, at physical sites, and in printed sources can also facilitate the development of observational skills. Studies in aesthetics, art criticism, art history, and visual culture foster students' observation and visual literacy skills because they place students in the habit of slowing down. Students learn to look carefully to notice details and find evidence—like STEAM professionals do. For example, while examining Katsushika Hokusai's (1760–1849) *Under the Wave at Kanagawa* (*The Great Wave*) (1831), students can discover how Hokusai strengthened his observational skills by drawing over the course of his lifetime (Thomas, 2015). He looked closely and paid attention to details to build a visual database in his mind for drawing. This practice enabled him to create *The Great Wave* as a convincing artwork. Having no experience with the actual wave himself, Hokusai most likely learned about the frightening event from sailors' detailed descriptions (Cartwright & Nakamura, 2009).

1.1.3 Imagination and Wonderment

Imagination and wonderment invite artists and designers to imagine new possibilities and feel inspired. Imagination manifests itself as mental constructions born of past experiences produced without the benefit of external sensory data input. Artists and designers nourish their imaginations by asking why and what if questions, experimenting, and engaging in creative play. Imagination can lead to innovative ideas, solutions, and products. Csikszentmihalyi (1997/2013) clarified: "Creative individuals alternate between imagination and fantasy at one end and a rooted sense of reality at the other" (p. 63).

The imaginative process can be intrinsically rewarding and generate heightened states of wonder. When artists and designers become intrinsically excited about their work, their products can also spark imagination and wonderment in others. For example, George Lucas applied his imagination to create the three *Star Wars* trilogies: the 1977–1983 original; the 1999–2005 prequel; and the 2015–2019 sequel. Lucas worked with STEAM specialists to bring his character designs, special effects, character personifications, and musical scores to life. Movie audiences felt a sense of awe and wonderment as they experienced the sensory appeal of the *Star Wars* trilogies onscreen. Lucas's innovative bluescreen compositing has also had a tremendous impact on teachers' and students' use of green–blue technologies for STEAM teaching and learning.

1.1.4 Persistence

Persistence centers on artists' and designers' abilities to persevere and be resilient—as they work through barriers and develop solutions. They revisit ideas, revise their work, fine-tune details, recheck hypotheses, and try things in entirely new ways when necessary. Being persistent and working through doubts and uncertainties, students can acquire determination and believe in their own potential and resolve. Persistence is identified in Anchor Standard 3 in the National Visual Arts Standards: "Refine and complete artistic work" and its essential question: "What role does persistence play in revising, refining, and developing work?" (National Coalition for Core Arts Standards [NCCAS], 2014).

Persistence is at the heart of the work of graphic designer Litha Soyizwapi and book arts engineer Sally Blakemore. Soyizwapi taught himself coding with limited resources. He developed *GauRider*, which became South Africa's No. 1 travel app in 2016, with the intent of serving commuters using the GauTrain rail and bus system. Soyizwapi persisted through failures as he became proficient in coding (TEDxTalks, 2014). He teaches others about his own experience with failure because it is a normal component of the creative process. Like all STEAM professionals, artists and designers need to understand how to become persistent and use failure to their advantage. Blakemore did not learn how to read until age nine due to reading barriers including dyslexia and synesthesia (personal communication, March 17, 2022). Loving picture books, she rotated them in different angles to process text. She applies her learning experiences of turning books in all directions to create mass-produced 360° pop-up books—such as her *NASCAR Popup Book: A Guide to the Sport* (Blakemore, 2009). Her designs demonstrate how students can acquire knowledge in different ways and spark their interest in books as learning tools.

1.1.5 Making Creative Connections

Making creative connections aids artists and designers in their ability to combine their life experience and existing knowledge to fuse concepts, beliefs, and values to form innovative and creative ideas. The National Visual Arts Standards (NCCAS, 2014) emphasize the importance of creative connections by incorporating the act of connecting into their framework with Anchor Standard 10: "Synthesize and relate knowledge and personal experiences to make art" and 11: "Relate artistic ideas and works with societal, cultural, and historical context to deepen understanding." Producing creative connections teaches students how to form genuine associations among STEAM disciplines. Students learn to select from resources, including data sets, materials, equipment, media, and processes to bring their ideas to fruition. These practices strengthen students' conceptual knowledge because they learn to identify how interrelationships foster deep understanding (Krathwohl, 2002). Individual disciplines need not be understood in isolation.

Reinforcing the concepts of creative connections, I identified three themes in my STEAM book: (a) sites of engagement; (b) art, design, and compiled data; and (c) going places (Sickler-Voigt, 2023). They connect seemingly different artists from a variety of times and geographical locations to assist teachers and students in melding themes in art and design with STEAM topics. For example, I explain how built environment designs including Antoni Gaudi's (1852–1926) *La Sagrada Familia* (1882–present) and Violet Oakley's (1874–1961) *Pennsylvania Capitol Murals* (1902–1927) have been informed by compiled data that have shaped societal understanding of their production and preservation.

1.2 *Designing Interdisciplinary and Transdisciplinary Studies*

Understanding artistic behaviors establishes a foundation for designing authentic interdisciplinary and transdisciplinary STEAM studies (Sickler-Voigt, 2023). Interdisciplinary studies involve two or more integrated disciplines equally informing one another using real-world applications (Huser et al., 2020; Liao, 2016; Mejias et al., 2021). By learning how to navigate the standards of the various disciplines of STEAM and identify how they can align when planning instruction (Costantino, 2018), educators have the tools they need to form logical connections among the STEAM disciplines and can teach students how practitioners of those disciplines may use different terminology when addressing similar topics. Integrating standards in the curriculum facilitates students' learning about the interrelationships of STEAM disciplines and augments their academic vocabularies.

Transdisciplinary studies (Huser, 2020; Liao, 2016; Mejias et al., 2021; Perales & Aróstegui, 2021; Sickler-Voigt, 2023) expand beyond interdisciplinary studies in that they holistically connect multiple STEAM subjects and are ideal for school-wide and community-based projects. Disciplinary boundaries are fully dissolved. Transdisciplinary studies are best achieved when professionals—including teachers in diverse disciplines and STEAM community partners—collaborate to share their expertise and augment possibilities. They integrate problem-based learning tasks built on authentic planning, instruction, and assessment to foster transdisciplinary STEAM studies that invite students to experiment, brainstorm ideas, and apply artistic behaviors as they identify creative solutions to problems (Quigley & Herro, 2019). Working together, students construct new knowledge that transcends existing possibilities and produce innovative results (Huser et al., 2020; Liao, 2016).

1.3 *Amplifying Teacher and Student Choice*

A choice-based STEAM curriculum offers continuous opportunities for teachers and students to make personal choices from among learning tasks, concepts, big ideas, materials, equipment, and subject matter (Douglas & Jaquith, 2018; Hetland et al., 2013; Sickler-Voigt, 2020). Rather than providing students with all the answers, teachers challenge them to explore ideas and seek solutions so they can make connections and gain confidence in their abilities as problem solvers and creators (Dubek et al., 2021). Peppler and Wohlwend (2018) emphasized how providing flexibility in students' use of learning materials and equipment meets the needs of their diversified learning preferences and provides them with greater engagement in curricular tasks and opportunities to produce successful learning outcomes.

In my book I emphasized authentic STEAM correlations, using rich narratives and choice-based learning tasks with ample flexibility to accommodate students' diverse ages and abilities, allotted instructional time, and available resources (Sickler-Voigt, 2023). I provided supplemental choice-based lessons, called STEAM Amplifiers, that build on artists' teachings using related STEAM topics. For example, I described how Deborah Butterfield created her sculpture *Three Sorrows* (2016) from tsunami debris and how its close examination can lead to studies on earthquakes, volcanology, and plastic marine debris. I explained how Jim Henson and his beloved Muppet Kermit the Frog (originated in 1955) can inspire additional lessons: (a) on neuroscience, the study of the human brain and how it influences people's abilities to set goals and change behaviors, drawing inspiration from Kermit's singing "The Rainbow Connection" while pondering his goals; (b) on laughter as students identify its many health benefits and pinpoint Kermit's humorous qualities; (c) on green scientific reporting that draws inspiration from Kermit's green coloring to teach about sustainable environmental practices; and (d) on financial literacy, which references Kermit the Frog's financial advice described in his autobiography *Before You Leap: A Frog's Eye View of Life's Greatest Lessons* (2006).

1.4 *Thriving through Hands-On Learning*

Hands-on learning is highly valued in the visual arts and authentic STEAM education (Costantino, 2018; Peppler & Wohlwend, 2018). In my book I designed instructional resources for educators to teach STEAM to PK–12 students using traditional art media and integrated technologies widely available in classrooms (Sickler-Voigt, 2023). Hands-on learning in STEAM education provides effective means to "take computation beyond the screen" (Peppler & Wohlwend, 2018, p. 89) and engage students in critical making practices that fuse higher-order thinking skills with the act of manipulating materials to generate new knowledge and solve creative problems (Costantino, 2018). I developed hands-on lessons that invite students to design architectural models, plein-air artworks, interactive performances, books, and puppets.

Wanting to model best practices for creating STEAM-themed artworks inspired by content, I collaborated with preservice art educator Paige Brenner as equal partners to create an original tunnel book for the cover artwork of my book. I also wanted to model how educators at different stages of their careers acquire technical skills so that they can teach them to their students. Brenner and I had experience creating handmade tunnel books and wanted to broaden our technological skills at our university's maker space. Planning our tunnel book's design, we analyzed characteristics of the artworks featured in the book and the content of the STEAM Amplifiers to create a multimedia artwork encapsulating the book's teachings. I began the process by designing its STEAM logo using the Procreate app and a stylus (Figure 2.1). Next, Brenner and I participated in tutorials to learn how to use a laser cutter engraving machine to construct the wood exterior of our tunnel book in the form of a vintage television set in response to Nam June Paik's (1995) video art installation *Electronic Superhighway: Continental U.S., Alaska, Hawaii* and its 336 television sets. We also cut the tunnel book's five layers of watercolor paper, called series, on the same laser cutter and then hand painted them to produce a vivid landscape. The serene landscape—complete with horses—acknowledges Franz Marc's (1911–1913) paintings of animals living in harmony with nature. A bright painted sunshine is analogous to Alma Thomas's solar paintings.

After additional maker space training, we produced 3D-printed wind turbines to symbolize Jenova Chen and Kellee Santiago's (2007) video game *Flower* and its teachings on renewable energy. Our final training was on LED lighting so that we could augment the book's aesthetic appeal with an integrated technology. Colorful, changing patterns pay homage to Janet Echelman's (2015–2020) LED-lit art installations and the benefits of LED over incandescent lighting. During a family visit my father Richard Sickler joined in the collaborative spirit of the design of our tunnel book and assisted us in assembling its many pieces and constructing a model bridge to hold a handmade sculpture of a high-speed train Brenner created—referencing Litha

FIGURE 2.1 This STEAM logo includes visual icons representing the content of *STEAM Teaching and Learning Through the Arts and Design: A Practical Guide for PK–12 Educators* (Sickler-Voigt, 2023). Its design symbolically unifies the STEAM disciplines and represents the value of interdisciplinary and transdisciplinary studies. © Debrah Sickler-Voigt

Soyizwapi's travel app and the importance of using sustainable energy to protect biodiverse environments.

Having completed our tunnel book, we reflected on our learning experience. We effectively applied each of the book's featured artistic behaviors—(a) idea development, (b) observation, (c) imagination and wonderment, (d) persistence, and (e) making creative connections. Idea development and observational studies flourished as we sought inspiration from the artists featured in the book and the STEAM Amplifiers. We transformed our ideas into a mixed media tunnel book. Despite initially lacking some technical and engineering skills, we were persistent and imaginative in what we wanted our design to become. Throughout the process we relied on guidance from knowledgeable partners. Maker space specialists trained and guided us to overcome technological barriers as we transformed our digital designs into laser-cut forms. Both experts in wood design, Richard Sickler and teaching artist Stephan Michelet-to-Blouin gave us directions on wood staining and construction techniques. Sickler assisted us in reengineering the tunnel book because its laser-cut slots, designed to hold the individual tunnel book watercolor series in place, were too thin and misaligned with the frame of the box. To solve this problem, we added tabs to each of the series to secure them in place and lend the box stability. We determined that lighting the tunnel book from its base rather than its interior top cover was the most effective way to hide the LED hardware. Our hands-on design processes greatly enriched our technical product and made the assembly of the tunnel book possible. The LED lighting added wonderment and variety to our design and inspired our development of a video that showcases the dynamic qualities of the tunnel book (Sickler & Sickler-Voigt, 2023). Ultimately, we achieved our goals and created a completed artwork that felt even more exhilarating than we initially imagined (Figures 2.2 and 2.3).

FIGURE 2.2
This photo of the tunnel book shown with white LED lighting showcases its removable hand-assembled rabbit ears. © Debrah Sickler-Voigt, Paige Brenner, and Richard Sickler
PHOTO: RICHARD SICKLER.

FIGURE 2.3 This photo collage shows the tunnel book with its range of LED hues. Its associated video illustrates its dynamic multimedia qualities (see Sickler & Sickler-Voigt, 2023). © Debrah Sickler-Voigt, Paige Brenner, and Richard Sickler PHOTO: RICHARD SICKLER

2 Implications and Conclusion

In this chapter I presented a model of authentic STEAM learning, encompassing four components: (a) artistic behaviors, (b) interdisciplinary and transdisciplinary studies, (c) teacher and student choice, and (d) hands-on learning (Sickler-Voigt, 2023). This model begins with STEAM education basics and illustrates authentic integration of the arts as coequal disciplines taught in context, using a range of their standards and vocabularies to avoid misinterpretation and stereotypes. When grounded in authentic planning, instruction, and assessment methods, educators are well-equipped to teach interdisciplinary STEAM lessons and can foster transdisciplinary STEAM experiences by collaborating with expert partners specializing in various STEAM disciplines.

The infinite possibilities of STEAM offer exciting pathways to implement interdisciplinary and transdisciplinary STEAM studies. I invite educators to use the instructional resources in *STEAM Teaching and Learning Through the Arts and Design* (Sickler-Voigt, 2023) as a foundation to integrate additional artists, designers, and artistic behaviors to suit their particular teaching interests and students' learning needs (see *Companion Websites*, 2023). The combined practices I have presented make the PK-12 STEAM curriculum feel rewarding to teach and motivate students to learn. Through sustained learning opportunities, students become adept in employing artistic behaviors and mindsets that they can call upon as lifelong learners and the creative STEAM innovators of the future.

References

Allina, B. (2018). The development of STEAM educational policy to promote student creativity and social empowerment. *Arts Education Policy Review, 119*(2), 77–87. https://doi.org/10.1080/10632913.2017.1296392

Blakemore, S. (2009). *NASCAR pop-up book: A guide to the sport* (NASCAR library collection). Gibbs Smith.

Cartwright, J., & Nakamura, H. (2009, February 25). What kind of a wave is Hokusai's *Great Wave off Kanagawa*? *Notes and Records of the Royal Society, 63*(2), 119–135. https://doi.org/10.1098/rsnr.2007.0039

Companion websites for Sickler-Voigt. (2023). https://routledgetextbooks.com/textbooks/_author/sickler-voigt/

Costantino, T. (2018). STEAM by another name: Transdisciplinary practice in art and design education. *Arts Education Policy Review, 119*(2), 100–106. https://doi.org/10.1080/10632913.2017.1292973

Csikszentmihalyi, M. (2013). *Creativity: Flow and the psychology of discovery and invention.* Harper Perennial. (Original work published 1997)

Douglas, K. M., & Jaquith, D. B. (2018). *Engaging learners through artmaking: Choice-based art education in the classroom (TAB)* (2nd ed.). Teacher's College Press.

Dubek, M., DeLuca, C., & Rickey, N. (2021). Unlocking the potential of STEAM education: How exemplary teachers navigate assessment challenges. *Journal of Educational Research, 114*(6), 513–525. https://doi-org.ezproxy.mtsu.edu/10.1080/00220671.2021.1990002

Eisner, E. W. (1998). *The enlightened eye: Qualitative inquiry and the enhancement of educational practice.* Merrill.

ElSayary, A. (2021). Transdisciplinary STEAM curriculum design and authentic assessment in online learning: A model of cognitive, psychomotor, and affective domains. *Journal of Turkish Science Education, 18*(3), 493–511. https://orcid.org/0000-0002-5554-0069

Every Student Succeeds Act, 20 U.S.C. § 6301 (2015). https://www.congress.gov/114/plaws/publ95/PLAW-114publ95.pdf

Hetland, L., Winner E., Veenema, S., & Sheridan, K. (2013). *Studio thinking 2: The real benefits of visual arts education.* Teachers College Press.

Huser, J., et al. (2020). *STEAM and the role of the arts in STEM.* State Education Agency Directors of Arts Education.

Kermit the Frog. (2006). *Before you leap: A frog's eye view of life's greatest lessons.* Disney.

Krathwohl, D. R. (2002). A revision of Bloom's Taxonomy: An overview. *Theory into Practice, 41*(4), 212–218.

Liao, C. (2016). From interdisciplinary to transdisciplinary: An arts-integrated approach to STEAM. *Art Education, 66*(6), 44–49. https://doi.org/10.1080/00043125.2016.1224873

MacDonald, A., Wise, K., Tregloan, K., Fountain, W., Wallis, L., & Holmstrom, N. (2020). Designing STEAM education: Fostering relationality through design-led disruption. *International Journal of Art & Design Education, 39*, 227–241. https://doi-org.ezproxy.mtsu.edu/10.1111/jade.12258

Mejias S., Thompson, N., Sedas, R.M., Rosin, M., Soep, E., Peppler, K., Roche, J., Wong, J., Hurley, M., Bell, P., & Bevan, B. (2021). The trouble with STEAM and why we use it anyway. *Science Education, 105*(2), 209–231. https://doi.org/10.1002/sce.21605

National Coalition for Core Arts Standards (NCCAS). (2014). *National Core Arts Standards*. https://www.nationalartsstandards.org

Peppler, K., & Wohlwend, K. (2018). Theorizing the nexus of STEAM practice. *Arts Education Policy Review, 119*(2), 88–99. https://doi.org/10.1080/10632913.2017.1316331

Perales, F. J., & Aróstegui, J. L. (2021). The STEAM approach: Implementation and educational, social and economic consequences. *Arts Education Policy Review*. https://doi.org/10.1080/10632913.2021.1974997

Quigley, C. F., & Herro, D. (2019). *An educator's guide to STEAM: Engaging students using real-world problems*. Teachers College Press.

Rhode Island School of Design. (n.d.). *STEAM*. https://www.risd.edu/steam

Sickler, R., & Sickler-Voigt, D. C. (2023, March 9). *STEAM teaching & learning through the arts and design: Book cover trailer* [Video]. YouTube. https://www.youtube.com/watch?v=9E_7JNdW4JE&t=1sA

Sickler-Voigt, D. C. (2020). *Teaching and learning in art education: Cultivating students' potential from pre-K through high school*. Routledge.

Sickler-Voigt, D. C. (2023). *STEAM teaching and learning through the arts and design: A practical guide for PK–12 educators*. Routledge.

Smithsonian Institution Archives of American Art. (2019, August 13). *Autobiographical writings by Alma Thomas, circa 1960s–circa* 1970s. https://edan.si.edu/transcription/pdf_files/17790.pdf

TEDxTalks. (2014). *How I taught myself to code: Litha Soyizwapi at TEDxSoweto* [Video]. YouTube. https://www.youtube.com/watch?v=nmZR_JsLCfg

Thomas, S. E. (2015). *Hokusai*. MFA Publications.

Topalska, R. (2021). STEAM education in the view of the Bulgarian teacher. *TEM Journal, 10*(4), 1822–1827. https://doi.org/10.18421/TEM104-45

CHAPTER 3

Challenges and Prospects for STEAM Education in Japan

Considerations from the Perspective of Art Education

Isao Ohashi

Abstract

Because art is an essential subject in STEAM education, the analysis of theories and practices of art education and the ways they influence or are integrated into it is necessary to understand STEAM education in Japan. To this end, I analyze the role of art and art education and its relevance to STEAM education, emphasizing its profound significance in fostering creativity and innovation, generating larger probing questions, facilitating the search of intrinsic value in life, enabling self-actualization, and strengthening comprehensive humanistic learning. Exemplified by Japanese students' STEAM projects and learning, the educational theories explained below include artistic play, education through art, expressive behavior, creative attitude, and self-actualization. Ultimately, I aim to offer pedagogical concepts and examples derived from the Japanese educational setting to accentuate how art integration can serve as a catalyst to augment students' STEAM educational experiences.

Keywords

art education – STEAM projects – self-actualization – artistic play – education through art

1 Introduction

In his exploration of the link between creativity and human behavior, Abraham Maslow (1949), a prominent American psychologist, identified two distinct components: coping and expressive behaviors. Coping behavior involves external determinants, emphasizing the manner in which individuals respond to tasks by seeking correct answers without spontaneous engagement. He argued that modern society values the pragmatic and result-oriented characteristics

 | DOI:10.1163/9789004714748_004

associated with coping behaviors. By contrast, expressive behavior is influenced by inner qualities such as "beauty, art, fun, play, wonder, awe, joy, love, [and] happiness" (Maslow, 1949, p. 261). Unfortunately, expressive behaviors are often considered useless reactions and tend to be overlooked.

Although a result-oriented problem-solving approach can effectively meet many societal needs by following predetermined structures, Maslow (1970) cautioned against exclusively relying on coping behaviors. He acknowledged that needs gratification, a motivating force behind coping behavior, is essential but highlighted the positive contribution of passion, an expressive behavior, in realizing an individual's full potential. Maslow called this phenomenon "self-actualization," a growth motivation utilizing inherent characteristics. Consequently, for learning to advance toward self-actualization, expressive behaviors driven by inner motivation are crucial, surpassing exclusive reliance on coping behaviors. Maslow (1971) underscored the significance of education that aligns with this perspective, emphasizing education through art as a means to achieve self-actualization. Expressive behaviors play a pivotal role in this context, allowing individuals to engage in spontaneous and uncriticized actions and creations. Thus, education through art becomes a pathway to unlocking one's full potential, fostering self-actualization through the encouragement of expressive behaviors aligned with inner drives.

Interestingly, the French anthropologist Claude Lévi-Strauss (1968) introduced the concept of bricolage, explaining how mythical thoughts operate through binary oppositions and structural relationships, such as life and death, nature and culture, or good and evil. Although mythical thoughts emphasize organized patterns from structural and cognitive aspects, the process of mythmaking itself can be seen as the work of bricolage, involving reusing whatever is at hand to cobble together an object to serve a new purpose. He further juxtaposed the problem-solving approaches of engineers, who work systematically and rationally, with bricoleurs, who improvise with spontaneity. Lévi-Strauss' argument underscores the notion that certain complex problems demand both systematic thinking akin to engineers and the creative use of materials characteristic of bricoleurs.

Tackling today's societal challenges like division, discrimination, international conflicts, and environmental crises requires a multifaceted approach. Individuals must look beyond their personal interests and ideological differences and actively address these issues. Doing so involves blending the problem-solving skills of engineers who approach challenges methodically with the creativity of bricoleurs who find innovative ways to use available resources.

Lévi-Strauss's (1968) concept of the artist as bricoleur is evident in the learning approach known as *zoukei asobi* (artistic play), an educational concept and activity in Japan that demonstrates the intricate process of the creativity involved in bricolage through artistic meaning making (Kirita, 2017). The artistic play approach encourages students to use available resources creatively to solve a particular problem. Children employ various semiotic manipulations, such as substitution, permutation, addition, subtraction, and equivalence, to engage with the activity; furthermore, the concept of STEAM, which incorporates Art into STEM fields, helps foster creativity for progress and sustainable knowledge.

Educational curriculum and activities have been standardized by the government in Japan for all subjects. STEAM education is mentioned only in a cross-curricular learning activity called "integrated inquiry time" in upper secondary schools. The application and interpretation of the "A" in STEAM predominantly revolves around subjects related to the liberal arts. Unfortunately, the educational significance of exploring unanswered questions creatively, which lies at the heart of art, is often overlooked in these examples. At most, many teachers incorporate the scientific process into their lessons, involving activities like researching previous studies, conducting experiments, developing mathematical models, and communicating findings through written and visual presentations. Although incorporating components of art and culture, such as having students participating in dyeing and weaving, this cross-curricular learning approach remains largely STEM-focused in its process and purpose.

In this chapter I highlight the role of art (the "A") in STEAM education, emphasizing its profound significance in fostering creativity and innovation, generating broader probing questions, facilitating the search for intrinsic value in life, enabling self-actualization, and strengthening comprehensive humanistic learning. This sets the foundation for understanding Japanese art education and STEAM education. By analyzing Japanese students' projects and learning, I illuminate educational concepts, including artistic play, education through art, expressive behavior, creative attitude, and self-actualization, deemed vital in art and STEAM education. My ultimate goal is to offer pedagogical concepts and examples derived from the Japanese educational setting to enable an understanding of how art integration can serve as a catalyst to enrich students' educational experiences and comprehension of complex concepts.

2 Significance of Art in STEAM Education

In a technologically advanced society, design plays a pivotal role, exerting its influence over consumer decisions by leveraging aesthetic allure. Maeda (2012) highlighted the inadequacy of functionality: Individuals now seek products and experiences that align with their values. This significant shift in consumer preferences underscores people's longing for deeper connection to the intrinsic value of life, no longer satisfied with mere material possessions. In this context he theorized that the true innovation needed to satisfy this longing would spring from the realm of art. According to Maeda, contemporary artists are dedicated to asking questions about intention, meaning, and truth, contributing to a deeper exploration and actualization of the inherent value in human life and society. Instead of confining themselves to the traditional boundaries of the art world, Maeda held that artists embracing the STEAM paradigm are particularly avid in asking broad, complex probing questions. His argument echoed Maslow's (1970; 1971) concept of self-actualization and Lévi-Strauss' (1968) notion of bricolage. Indeed, artists commonly seek self-actualization like bricoleurs, giving priority to authenticity and integrity by aligning with their inner convictions, all the while remaining unaffected by external judgments. This disposition resonates with art education, which prioritizes self-expression, aiming to encourage individuals to communicate and cultivate their distinct voices through artistic pursuits. Thus, Maeda's insights underscored the profound interplay between art, individual authenticity, and societal innovation, idealizing the relevance of how creativity shapes contemporary culture. STEM education alone cannot fully nurture these unique characteristics.

To differentiate designers from artists, Maeda (2012) stated: "Designers create solutions—the products and services that propel us forward. But artists create *questions*—the deep probing of purpose and meaning that sometimes takes us backward and sideways to reveal which way *forward* actually is" (para 8). In other words, through their expressive activities, artists do not solve problems that already exist; instead, they introduce challenges that have yet to be recognized by anyone. For instance, Marcel Duchamp made a daring attempt to showcase a fully assembled urinal unit, titled *Fountain* (1917), at New York's newly formed group exhibition organized by the Society of Independent Artists (Mendelssohn, 2019). He submitted this work as R. Mutt, playing on the name of the J. L. Mott Iron Works where he had purchased the toilet bowl. In principle, as the antithesis of the authoritarianism prevalent up to that time, the Society of Independent Artists was supposed to exhibit all entries; however, the committee rejected *Fountain*, arguing that it did not qualify as art.

This incident sparked two critical concerns, leading to notable shifts in the art world. The first concern was what qualifies as art, prompting a reevaluation of the definition of art. Does modifying an existing ready-made object make it art? This led to a greater focus on originality and ideas rather than just technical skill in the process of creating a work of art. The second concern dealt with the persistence of authoritarianism in the art world. Despite claims of breaking away from authoritarian practices, the exhibition organizers still imposed criteria on which works could be displayed, contradicting their supposed ideal of freedom. Duchamp intentionally challenged such authoritarian practices, promoting ongoing debates over underlining authoritarian control.

Regarding reasons that artists create, however, they do not necessarily aim to change others or society through their artistic visions. Instead, they often turn inward, reflecting on what they personally want to accomplish, using visual elements to communicate with others. From this self-exploration process their creations come to life; that is, the artwork resolves artists' inquiries or a significant point of view along their journey. In the meantime, viewers are urged to ponder more questions when connecting with the artworks. This interconnected relationship shared by the artist, the artwork, and the audience underscores the educational value of artistic activities, emphasizing the importance of posing questions with multiple answers and seeking solutions through exploration. Only when the educational merits of these artistic pursuits are integrated into STEM education, can an educational approach that maximizes individual potential be shaped. Maeda (2012) stated that designers, whose creative processes are similar to artists, should possess the ability to formulate questions. He advocated for a shift from STEM to STEAM education to foster a more comprehensive learning and making approach. Today, the concept of design thinking + art thinking is not a division of designers and artists but instead a fundamental concept of STEAM design.

Many high schools in Japan have implemented STEAM education, but a notable gap exists in adapting the original concept proposed by Maeda (2012). Instead of embracing STEAM's holistic approach, many proponents in Japan primarily emphasize its connection to mathematical education. Artistic disciplines are often relegated to a supporting role without full integration; consequently, people mistakenly believe that simply combining math and art subjects constitutes STEAM education.

STEAM education holds considerable significance as it aims to cultivate individuals who can effectively harness and synergize both their understanding and application of the natural order of the external world and the imagination, creativity, sensitivity, and aesthetic intuition that emerge from within. Although STEAM education in Japan needs further development, I have noticed many teachers endeavoring to implement STEAM activities and observed children exhibiting a natural tendency toward STEAM learning. Let me describe a scenario of two toddlers playing in a sandbox constructing a large sand pile. One toddler (Child A) pours buckets of water onto the top of the pile, causing it to collapse because of the sudden influx of water. Undeterred, another toddler (Child B) attempts to rebuild the pile, only for Child A to pour water repeatedly, leading to further collapses. After several iterations, Child B proposes a new approach as a resolution: digging a trench around the sand pile to contain pouring water. Dissolving conflicts with an opportunity for collaborative work to move forward, they transform their competitive play into a simulated river, meandering, widening, and narrowing the pathway to mimic natural landscapes. In doing so, they harness the physical elements around them and experience the joy of creating an idealized world together.

3 Theoretical Components Needed to Augment STEAM Education

Because the time allocated for art class at schools in Japan is limited, teachers tend to focus on how to produce artwork effectively. Providing opportunities for students to achieve self-actualization through enthusiastic participation proves to be challenging, students often struggling to gain comprehensive knowledge and needed skillsets for both materials and tools. This educational process, perpetuating the coping behavior, does not differ significantly from top-down linear models that aim solely for specific outcomes. Considering this setback and art as one of the subjects in STEAM, I propose two art educational theories—artistic play and education through art—to augment STEAM education and to reach the educational goal of self-actualization.

3.1 *Artistic Play in STEAM Education*

Zoukei asobi (artistic play), a key component in the arts and crafts curriculum in Japanese elementary schools, can be applied to STEAM education. I vividly recall an instance watching a five-year-old playing with a milk carton. The child soon discovered the paper's notable elasticity. This led the child to creatively design a contraption by rolling up pieces of paper to form a ball and cutting out a paper plate from the milk carton as a bouncer. The child then underwent a series of trial-and-error attempts to fashion an effective plate spring. With unwavering determination, the child successfully accomplished the task.

Prevalent in early childhood education in Japan, *zoukei asobi* was designed to foster a child's interest in exploring various materials and the sheer joy of hands-on studio practices. In addition to focusing on producing artworks through expressive activities like painting, 3D modeling, design, and crafts, *zoukei asobi* emphasizes engaging children through play to experience creating with spontaneity and intuition. Children's enthusiasm toward play promotes creative expression in the arts. In educational activities emphasizing the importance of play, children act as bricoleurs, applying, presenting, and acquiring knowledge and skills experientially and creatively. Through active playing, they can fully tap into their potential and move toward self-actualization.

In *zoukei asobi*, the teacher does not impose learning activities with specific outcomes; instead, the activities focus on exploring materials—often items recycled from daily life, seen as treasures by children. Through imaginative play, familiar surroundings can be transformed into extraordinary settings. As children manipulate materials and become excited about using their unique characteristics, they delve into their potential. Alternatively, they might experience setbacks but learn to leverage available materials in their surroundings to solve problems. This learning-by-playing attitude fosters an inquisitive attitude akin to a scientist seeking truth while simultaneously providing moments of inspiration, resembling the trial-and-error process an artist undergoes on the path to self-actualization.

Through *zoukei asobi,* the child who played with a milk carton experienced a learning process that exemplifies the essence of STEAM education. In contrast to activities where the teacher presents a problem for the child to solve, this child gained insights that extend beyond conventional learning. This child's learning experience characterizes the integrated nature of STEAM, where distinctions between science and art dissolve naturally in play. Consequently, in the art component (the "A") in STEAM, akin to the approach in *zoukei asobi*, the learning process is not merely a problem-solving exercise but a journey toward trial-and-error discovery and self-actualization through a self-motivated, creative learning mindset. This involves a dynamic interplay of making while thinking and thinking while making—an iterative process in which value is creatively realized through the continual exchange of action and thought.

4 Education through Art in STEAM Education

Maslow (1971) stated that education through art is another component to be incorporated into STEAM education. He asserted that similarities in "the concept of creativeness and the concept of the healthy, self-actualizing, fully human person [are evident; therefore,] education through art may be especially important not so much for turning out artists or art products, as turning out better people" (p. 57). STEAM education can be perceived as an education, less for reproducing artists or scientists or artistic or scientific products per se but for nurturing a personality that fully harnesses its potential toward self-actualization through activities marked by expression and appreciation. Encouraging overarching, forward, and future thinking, not limiting education to a short-term, specialized, or rote training, this conviction is inherent in artists and can also be extended to scientists and individuals across all disciplines who use creativity to drive innovation. Individuals so inclined tend to develop distinct qualities and abilities, including the ability to ask probing questions, create novel values, and tackle problems from a broad perspective. Education through art can, therefore, be a goal that we should aspire to cultivate in and through STEAM education.

I outline a learning structure based on the process of self-actualization in Figure 3.1 (Ohashi, 2021). As noted above, the emphasis on expressive behavior, rather than coping behavior, is crucial for learning because it aligns with the journey toward self-actualization. Coping behavior, characterized by excessive concern for others' opinions, hinders the emergence of expressive behavior. To guide learners toward the *creative attitude* as described by Maslow (1971), wherein they demonstrate creativity on the path to self-actualization, liberating them from coping behavior and encouraging them to embrace expressive behavior are essential. This process commences with becoming absorbed or, in other words, immersing oneself. When learners reach a state of absorption, in which external judgments, evaluations, and assigned tasks all fade into the background, they then are able to liberate themselves from coping behavior and gradually transition toward the self-motivated learning and discovery deemed vital in expressive behavior. This immersive state serves as a catalyst for fostering the creative attitude, a state Maslow called "creative furor" (p. 59).

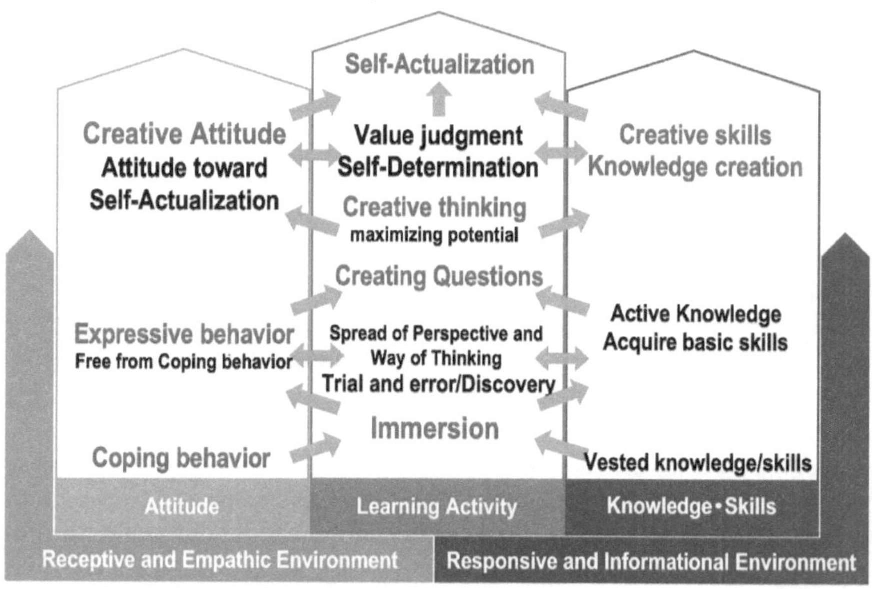

FIGURE 3.1 Learning structure as a process of self-actualization based on Ohashi (2021) and updated for this chapter

Indeed, teachers need to be aware that coping behavior can act as an inhibitor in the learning process. For instance, when students are deeply engrossed in painting, the state of immersion may be disrupted if they become aware of others' opinions. While immersion temporarily liberates them from coping behavior, without a supportive and understanding environment where worries about external judgments are alleviated, they may retreat into a coping mindset.

Only in a receptive and empathetic environment that individuals can fully immerse themselves in the activity, paving the way for expressive behavior and subsequently opening the door to creative attitudes. This environment is characterized by a sense of safety and acceptance, where students feel confident that their expressions will not be dismissed or ridiculed. Essentially, it is a human-centered rather than solely a mechanical environment. Both the teacher–student relationship and the student–student dynamic play crucial roles in creating and maintaining such an atmosphere. The most important priority for teachers is to cultivate an inclusive environment within all aspects of school life where students feel safe to embrace and respect one another's individuality. For instance, assignments with only one correct answer often instill fear of making mistakes; however, in education through art, where exploration is encouraged, no answers are inherently wrong. Students experience only technical failures. Distinguishing between failure and a wrong answer is crucial. In a receptive and empathetic environment, failure is embraced as a natural part of the learning process. In this environment students are assured that failure and trial and error are integral to growth.

5 Students' STEAM Projects

Theories of *zoukei asobi* (artistic play) and education through art are embedded in Japanese art education. These theories have also been applied to support students' learning and the making of STEAM projects. In this section I describe students' learning processes, highlighting their probing questions, creative thinking, trial-and-error experiments, and STEAM aptitude.

5.1 *A Rabbit Born from a Piece of Grid Paper*

Figure 3.2 is the artwork of a third-grade elementary student, titled "Rabbits With a Sword and a Fishing Rod." Although it may appear to be a typical paper craft at first glance, the student received no guidance on using grid paper or crafting 3D objects from paper. The inception of this *zoukei asobi* project stemmed from a child's accidental discovery of a piece of grid paper. I observed the creative and problem-solving process of this child:

FIGURE 3.2 Rabbits with a sword and a fishing rod

- After studying the grid paper for some time, the child drew the eyes, mouth, and other facial features based on the grid. She pondered, "What can I do with this?" and then conceived the idea of using rectangles along the grid to create her image.
- This child created a three-dimensional head by carefully cutting and folding along the designated grid lines. She asked herself, "What can I do?" and in response, she proceeded to the next actions in which she cut and folded and assembled various shapes.
- Following the grid as a guide, she cut, folded, assembled, opened, and reassembled the body, limbs, a sword, and a fishing rod. She continually made, remade, and created, enhancing her thoughts and actions and ultimately finding a new purpose to work with the materials.
- The child further attempted to make the paper doll stand but struggled to maintain its stability. I observed that the center of gravity was high, leading to poor stability. Interestingly, the child seemed to prefer this precarious balance

and was reluctant to make it more stable. After numerous trials, she eventually achieved success by creating a flat base under the doll's feet and securing it in place with glue. She considered how to prevent the doll from toppling over without altering the image and devised her own solution to the problem.

For a third grader, creating a 3D object from a single sheet of paper by thinking logically and calculating with reference to a grid is not an easy task; however, I witnessed this child embracing the playful exploration of trial and error while tapping into her aesthetic preference, logical thinking, and problem-solving skills resembling STEAM learning. She invented her own approach and methods for bringing her vision to life. The crucial factor that led her down this inventive path was undoubtedly the supportive and empathetic environment, enabling her to fully engage with the challenges at hand. This detour represents a creative learning process characterized by curiosity and an innovative attitude.

5.2 *Mechanical Work Using Programming*

Figure 3.3 features a STEAM project made by fifth-grade elementary school students participating in the lesson titled Mechanical Work, a component of

FIGURE 3.3 Out! The crocodiles are coming! (If you fail in a game programmed with the Micro:bit, you will be attacked by an alligator.)

the art and handicraft curriculum. Mechanical Work involves activities that encourage students to imagine and create using devices and mechanisms. In this particular lesson, students used Micro:bit, a microcomputer system designed for programming education and widely adopted in schools globally. They employed programming to control a servomotor and produce various motions through a linkage and crank mechanism. Through a process of trial and error, students gained an understanding of how to control the servomotor and combine various mechanisms to achieve several types of motion. Subsequently, they explored how to program the desired movements. Once they had a clear vision of what they wished to express, they then used a range of materials to create artwork employing artistic thinking and skills.

I was impressed with how comfortably and naturally students learned to program using PCs and tablet PCs as tools, just like using scissors or glue as shown in Figure 3.4. The focus here is not solely on the acquisition of PC or programming skills but instead the holistic application of programming, thinking, and aesthetic consideration. This entails navigating trial and error to actualize one's thoughts or images, emphasizing the fusion of programming logic and creative expression. Efficient programming descriptions are essential to

FIGURE 3.4 The children used the PC as fluently as they used scissors and glue to give shape to their ideas

achieve optimal results, but recognizing that obtaining a single correct answer is not the primary objective of this activity is crucial.

Students engaged in self-reflection by asking themselves probing questions throughout this exercise. In fact, their STEAM journey began with a question that sparked curiosity about the function and mechanism of the task at hand. These questions were rooted in the anticipation that something intriguing could be achieved. Below are a student's questions and reflections as he embarked on creating the crocodile robot.

- The student encountered the Micro:bit system and found the concept of controlling diverse movements through programming intriguing. He expressed optimism that "something interesting could be done with it."
- Next, he delved into experimenting with the connection between the program and the movements generated by operating the PC. He engaged in trial and error with the program, making various changes while asking himself, "What kind of movement can I create with this?"
- In formulating the program, he experimented with various triggers to initiate motor movement. He pondered, "What kind of trigger would be enjoyable to make the motor move?" Consequently, he programmed a game in which the motor would activate when the player lost.
- The response of the motor to losing the game conjured an image of the powerful jaws of a crocodile, heightening his anticipation for bringing that image to life. Subsequently, he delved into exploring the movements and images he envisioned, addressing issues like what materials to use and how to assemble a large mouth to actualize the depiction of the crocodile biting the loser.

I noticed how this student's initial curiosity about the functionality of a tool evolved. From pondering what is achievable, a clearer vision emerged of what he desired to do and create. Throughout this process, aesthetic and sensory considerations gradually developed. Questions like "How can I enhance its beauty?" and "How can I convey the desired shapes, colors, and textures?" demonstrate his formative visual perspective, influencing choices regarding shapes, colors, and materials. Consequently, the student's creation serves as a response to his own inquiry and embodies the values he sought to realize. This iterative process of creativity directed toward self-actualization epitomizes the essence and objectives of STEAM education.

5.3 *Conveying the Japanese Aesthetic to Foreign Students*

Besides participating in expressive activities, activities in arts appreciation hold significant value in the field of art education and can support STEAM education. Many art educators in the past viewed learning art appreciation as incompatible with fostering hands-on creativity because it primarily focused on studying the stylistic history of art and learning about artists and their works. Today, however, it is recognized as a creative learning endeavor that encourages students to think critically; moreover, recognition of the educational benefits of appreciation activities in nurturing aesthetic sensitivity and intuition has grown over time.

Let us consider the aesthetic concept of *wabi-sabi* in Japan, a notion challenging even for native Japanese to grasp fully. *Wabi-sabi* is intricately tied to the traditional Japanese tea ceremony. Originally, the term *wabi* conveyed negative emotions like dreariness or loneliness, and *sabi* referred to the state of being old, decayed, dirty, or having lost its original beauty. In the context of the tea ceremony, however, *wabi* can denote simple and uncomplicated beauty, and *sabi* can signify tasteful beauty, resulting from natural aging, which cannot be deliberately created. In the following section, I describe an example of an art class in a Japanese junior high school, where students learned about the profound meaning of the Japanese aesthetic of *wabi-sabi*. In this class, students were tasked to explain the beauty of *wabi-sabi* to a foreign student in a way easy to understand. Students used pictures of tea rooms and tea bowls from the textbook as their teaching materials. Below are some responses provided by the students:

– The reason the bowl by Chojiro is not perfectly shaped is to take advantage of the beauty of nature and to value the spirit of the bowl over its material condition.
– *Wabi-sabi* involves feeling the beauty of nature as it is, like the bowl made by Chojiro, without trying to make it pretty or gorgeous.
– For instance, iron rusts and wood decays over time. *Wabi-sabi* is a mindset that acknowledges the natural progression of change and seeks to appreciate its beauty along the way.
– Similarly, when designing a space using limited wood on hand to build the most possible walls reflects the concept of *wabi*, where we embrace the imperfections and enjoy filling the gaps with our own heartfelt creativity.

Students gathered essential information from both literature and the Internet. With this information as their foundation, they grasped the concept of *wabi-sabi* in their own way. Furthermore, they attempted to explain it in simple terms, using as many words as necessary, while imagining themselves as foreign students with limited Japanese proficiency. In order to explain information to international students clearly, students had to research and sometimes substitute words with simpler Japanese equivalents. This process often revealed their own lack of understanding of *wabi-sabi*. This task prompted them to reflect on whether they truly grasped Japanese culture and aesthetics. The task of conveying information to others became an opportunity for self-evaluation of the authenticity of their own comprehension.

The teacher dimmed the lights in the classroom, creating a subdued atmosphere to facilitate discussion. Students could physically touch a tea bowl. This hands-on experience was very different from seeing tea bowls in the textbook or in the traditional tearoom setting. Students offered comments, such as the following: "Does the pattern of the tea bowl hold significance in the subdued lighting of a tearoom?"; "If it's difficult to grasp [the bowl], I'm worried about dropping it. Wouldn't a shape that naturally fits in the hand be preferable over a meticulously crafted one?"; and "I prefer lighting that seamlessly integrates into the room." This activity provided the students with an opportunity to truly comprehend the reasons behind the value Japanese place on dimly lit tea rooms and the rustic charm of pitch-black tea bowls. In addition, students reckoned that this activity encouraged them to "embrace the beauty of nature and prioritize the spiritual over the material," "appreciate nature's inherent beauty," "acknowledge and appreciate nature's transformations," and "find fulfillment with one's own emotions."

Students' comments eloquently capture the essence of traditional Japanese aesthetics. For example, Okakura (1906), one of the world's leading authorities on Japanese aesthetics, stated the following at the beginning of his book *The Book of Tea*:

> Teaism is a cult founded on the adoration of the beautiful among the sordid facts of everyday existence. It inculcates purity and harmony, the mystery of mutual charity, the romanticism of the social order. It is essentially a worship of the Imperfect, as it is a tender attempt to accomplish something impossible thing we know as life. (p. 1)

The essence of teaism, as pointed out by Okakura, is *worship of the imperfect*, a concept that the art teacher set out to help the students comprehend. In addition, Okakura discussed the origins of Japanese aesthetics, emphasizing a sense of humility standing before nature. Thus, the idea of the worship of the imperfect lies at the core of the belief that beauty resides in imperfection, requiring a mindset to appreciate it in a dimly lit environment where perfection is obscured and difficult to detect. This notion suggests that beauty is not bestowed upon humans but rather pursued by them, with the worship of the imperfect serving as the foundation of this concept.

To find what they needed to explain *wabi-sabi* to foreigners, students used their imagination, drew from personal experiences, and applied metaphors to

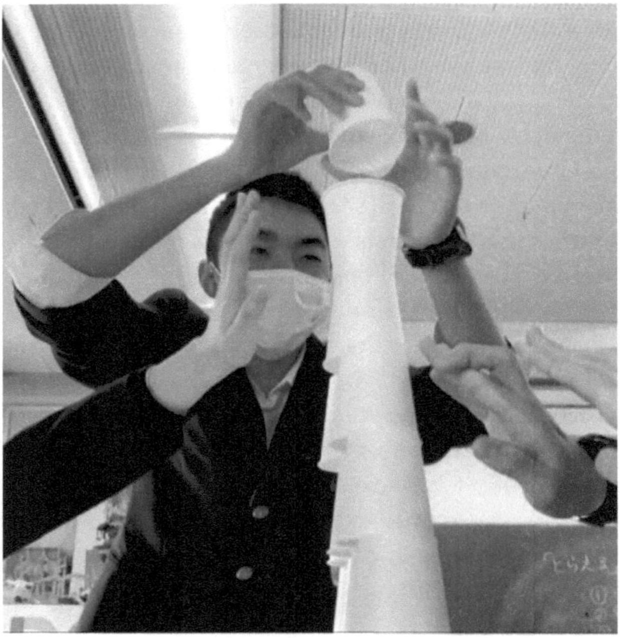

FIGURE 3.5 Paper Cup Tower Activity, conceived by Dr. Tetsuo Kiyota and practiced by Ms. Hiroko Hanazato at Nakano 5th Junior High School, Tokyo

convey its meaning. This process embodied artistic thinking. The study of art served as a platform for posing questions and resolving them creatively. Students were urged to harness their full potential to actualize the values they deemed significant. I believe that only with this creative mindset as the foundation of learning can the essence of STEAM education be truly realized.

5.4 *Team Collaboration Transforms Individuals and Groups*

An activity one might call "Let's make a beautiful tower by stacking paper cups" (see Figure 3.5) involved team-based learning. Described by Hanazato (2022, pp. 187–189), this activity surpassed a mere tower-building competition. Its goal was to foster collaboration among team members with various disabilities. For instance, restrictions included using only one hand, remaining seated, wearing a blindfold to hinder sight, or limiting the ability to speak.

How could students effectively communicate their goals and visions with members who were blind or nonverbal? Moreover, how did they navigate the frustration of desiring to accomplish something hindered by their disability? These challenges initially posed significant obstacles; however, the students successfully surmounted these difficulties by uniting around a shared aesthetic value: the aspiration to construct a beautiful tower collaboratively, employing their own hands and working alongside their friends.

As they strove to embody this aesthetic value, their initial limitations evolved into a collective strength through collaboration. Throughout the tower-building process, as they prevailed over various challenges, the students came to realize that each person possessed unique strengths and weaknesses. This realization underscored the transformative influence of art (the "A" in STEAM) in uniting them, ultimately harnessing their collective strength.

6 Final Reflection

In the final scene of Spielberg's (2002) movie *A.I. Artificial Intelligence*, robots created by humans govern the Earth, where humans have become extinct. These robots conduct research to comprehend the humans who crafted them utilizing information from memory media salvaged from the era when they coexisted with humans. Despite surpassing humans in every conceivable way, one aspect eludes their understanding. The robots, despite their advanced mastery of science and technology, grapple with the lifelong human practice of "art," a pursuit that appears by rational standards to be of no practical utility. Essentially, the film suggests that no matter how sophisticated a robot becomes, it cannot fully emulate humanity because of its inability to grasp the essence of art.

Certainly, the ongoing progress of science and technology will be instrumental in our future, yet recognizing that, we must not forget that the foundational soil that yields these fruits is crucial for immediate benefits. Indeed, genuine scientists and engineers have always embraced aesthetic sensibility and intuition. Science, technology, and art share a common ground from which the fundamental question arises: "What defines a human being?" Humans inherently appreciate the equal value of beauty and knowledge, and STEAM education aligns with this fundamental aspect of human nature. Within the actual educational landscape, however, the influence of rationality, efficiency, and economy persists, hindering a broader understanding of the role of art in education and its educational value. The field of art education must liberate itself from the grip of outdated approaches. A meaningful step forward involves sharing effective educational practices globally.

In this chapter, I offer pedagogical theories and examples supporting art education and STEAM education in Japan. In the Japanese education system, all subjects including the arts are mandatory in elementary and junior high schools. In many elementary schools, classroom teachers handle multiple subjects, including mathematics, language arts, social studies, the arts, and physical education. This arrangement facilitates not only specific subject-based learning but also allows an interdisciplinary approach that transcends traditional disciplinary boundaries. Moreover, Japanese official standards of the arts and crafts curriculum do not mandate students to create impressive artworks or solve problems imposed on them unilaterally. The teachers highlighted in this chapter actively engaged in art education often by combining what they taught with other academic subjects and occasionally by prompting students to confront environmental and social issues, encouraging them to formulate their own questions and solutions. Although STEAM education in Japan is still a new phenomenon, Japanese institutional features can pave the way for teachers to employ the concepts of *zoukei asobi* (artistic play) and education through art to enhance STEAM activities, and I believe all together they can form an ideal foundation for advancing STEAM education.

References

Hanazato, H. (2022). Kodomotachi ga toraeru asobi [Play captured by children]. In I. Ohashi (Ed.), *Bijyutukyouiku to kodomorikai* [*Art education and understanding children*] (pp. 182–189). Nihon Bunkyo Shuppan [Japan Bunkyo Publishing].

Kirita, K. (2017). Understanding the creativity of bricolage in zoukei-asobi. In G. Barton & M. Baguley (Eds.), *The Palgrave handbook of global arts education* (pp. 513–532). Palgrave Macmillan. https://doi.org/10.1057/978-1-137-55585-4_31

Levi-Strauss, C. (1968). *The savage mind.* University of Chicago Press.

Maeda, J. (2012). If design's no longer the killer differentiator, what is? *Wired.* https://www.wired.com/2012/09/so-if-designs-no-longer-the-killer-differentiator-what-is/

Maslow, A. H. (1949). The expressive component of behavior. *Psychological Review, 56*(5), 261–272. https://doi.org/10.1037/h0053630

Maslow, A. H. (1970). *Motivation and personality.* Harper & Row.

Maslow, A. H. (1971). *The farther reaches of human nature.* Viking Press.

Mendelssohn, J. (May 1, 2019). The essential Duchamp: An exotic radical who rejected the establishment. *The Conversation.* https://theconversation.com/the-essential-duchamp-an-exotic-radical-who-rejected-the-establishment-116157

Ohashi, I. (2021). Bijyutu ni yoru souzouteki-ningen no ikusei [Nurturing creative human beings through art]. In Nihon bijutsu kyouiku gakkai "70 shuunen kinen ronshuu" henshu iinkai [Japanese art education society "70th anniversary proceedings" editorial committee] (Eds.), *Mirai ni Tsunagu Bijyutukyouiku* [*Art Education for the Future*] (pp. 136–146). The Art Education Society of Japan.

Okakura, K. (1906). *The book of tea.* Dover Publications

Spielberg, S. (Director). (2002). *A.I. artificial intelligence.* Warner Bros. Pictures.

CHAPTER 4

Effect of STEAM Education on Schools and Universities in Türkiye

Vedat Özsoy

Abstract

This chapter highlights the efforts and limitations of STEM–STEAM education in Türkiye, where the Ministry of National Education has implemented two projects—eTwinning and Fostering STEAM Education in Schools (EDUSIMSTEAM)—for the enhancement of science, technology, engineering, art, and mathematics (STEAM) education. Although some research related to STEAM education has been conducted at universities, more studies have involved science, technology, engineering, and mathematics (STEM) education than art. In some private schools STEM education is provided under the name STEAM education, in which the design field is partially included, but the art field is not equitably represented. In this chapter I analyze the shortcomings of Türkiye's STEAM education and propose that the main reason for this is the scarcity of experts available to contribute to the "A" in STEAM education.

Keywords

art and STEAM – STEAM education in Türkiye – eTwinning – EDUSIMSTEAM

1 Introduction

One of the most influential educational paradigms of the 21st century is STEAM education, in which science, technology, engineering, art, and mathematics are integrated. Science, technology, engineering, and mathematics (STEM) and STEAM education models were tested and have been applied in public and private schools in Türkiye since the early 2000s. The 81-page STEM Report published by the Ministry of National Education (MoNE) (2016) recommended ways to develop STEM programs but included no examples of actual STEM application in schools; nevertheless, STEM education has captured attention in Türkiye. In my research I found that some private and public schools in

 | DOI:10.1163/9789004714748_005

Türkiye had already applied STEM education practices, and academic research on them has been conducted continuously (Ciftci et al., 2020; Durak et al., 2021; Yılmaz et al., 2018); however, more studies and reports on STEAM education in Türkiye are still needed to assure its advancement.

STEAM was promoted in the United States because researchers and educators believed STEM could not exist without art. STEAM education shapes learning by creating a structure in which science and technology are interpreted through or connected with engineering, art, and design. The integration of art into STEM plays an important role in the development of cognitive skills, such as expression, communication, creativity, imagination, observation, perception, and thought. It also improves problem solving and decision making for both students and teachers (Taylor, 2017). STEAM training aims to enable students to learn how to combine various disciplines with art and technology in a way that exhibits their creativity and technical abilities. STEAM education is not about spending less time on science to make room for art; it is about bringing these two disciplines together and looking in new ways (Subramaniam, 2021). STEAM is not a superficial integration or juxtaposition of art with science and technology. Rather, it is a cohesive project that sensibly combines diverse knowledge to improve the quality of education.

2 STEAM Education in Türkiye

To further understand STEAM education in Türkiye, in this section I highlight three types of studies that have been carried out there:
- studies conducted by the Ministry of National Education,
- studies featuring school-centered applications, and
- studies carried out at universities.

2.1 *Studies Conducted by the Ministry of National Education*

The MoNE carried out two projects: first, eTwinning and STEAM Education and second, Fostering STEAM Education in Schools (EDUSIMSTEAM). I summarize these two projects below.

2.1.1 eTwinning and STEAM Education

The European Union (EU) launched the eTwinning program as the flagship of the European Commission's Learning Program in 2005 (European Union [EU], 2022) so that educators could communicate, collaborate, develop, and share projects as well as engage in an exciting learning community in

Europe. The program provided a platform on which teachers, headmasters, librarians, and so on could effectively work together in a participating European school. eTwinning has been tightly integrated into Erasmus+, which has supported EU education, training, youth, and sports programs since 2014 (EU, 2020).

Türkiye participated in the eTwinning and STEAM Education project in 2009. An estimated 300,000 teachers were registered as of 2022, and more than 130,000 had participated in program activities through online education (Ministry of National Education [MoNE], 2022c). This shows that teachers were willing to engage in STEAM activities and to be part of a large EU educational community as equal educational partners, all without travel and bureaucratic paperwork. During this project, professional development opportunities, sharing groups, and discussion forums for all participating European teachers were created and successfully maintained.

eTwinning aimed to encourage member schools to cooperate by providing them with the necessary infrastructure to do so. The STEAM training provided online on the eTwinning platform by the MoNE's Education Information Network were divided into eight modules (MoNE, 2022c):
- STEAM Preliminary Survey,
- What Is the STEAM Education Approach?
- Why Has the STEAM Education Approach Gained Importance?
- How to Plan for STEAM Training,
- Preparing a STEAM Education Lesson Plan,
- STEAM Training Examples (see Figure 4.1),
- The STEAM Final Survey, and
- The STEAM Bibliography.

FIGURE 4.1 STEAM training examples: weather
PHOTOGRAPH: MONE (2022C)

A case study conducted by Gündüz Çetin and Gündoğdu (2022) revealed teachers' opinions about the eTwinning community. These researchers collected data through semistructured interviews with 10 teachers who worked in public schools during the 2020–2021 academic year and were active on the eTwinning platform. Notably, there were no art teachers among them. According to the case study, the teachers interviewed said that as a positive outcome of these practices, students' digital skills and learning motivation increased with eTwinning projects. Although the teachers met on the eTwinning platform in various ways, they stated that what motivated them most to stay on the platform was the development of their professional skills and the pleasure they experienced as a result of their success. In addition, the teachers said they had the opportunity to integrate more innovative approaches through eTwinning than through traditional teaching.

Again, notably, in this and similar studies on the eTwinning project, art teachers were not included among the working groups because they did not register for the eTwinning projects or participate in the related research studies (Avcı, 2021; Başaran et al., 2020). Thus, I cannot describe art teachers' experiences with the eTwinning projects. The lack of information about the importance of the arts in STEAM education in the eTwinning projects and how the arts are included in teaching practices raises some concerns.

2.1.2 Fostering STEAM Education in Schools (EDUSIMSTEAM)

Fostering STEAM Education in Schools (EDUSIMSTEAM) was another project conducted by the MoNE. This international project was launched in 2019 in Türkiye and supported by the European Commission under the Erasmus+ program. The main priority of the project was the development of innovative and multidisciplinary approaches in STEAM education (MoNE, 2022b).

The MoNE explained that EDUSIMSTEAM aimed to use information and communication technologies in innovative learning environments like STEAM education to transform developing and changing educational practices into multidisciplinary practices. EDUSIMSTEAM also aimed to share curricula and instruction, including STEAM learning scenarios based on real-life problems with all education stakeholders on online platforms. The innovative online platform, supported by various educational simulations of applications, was of considerable importance for the sustainability of the project (EU, 2021). The MoNE implemented the project with the contributions of 10 European partners. It also served objectives parallel to the practices of innovative and flexible learning environments supported by the European School Network, contributing to the dissemination of STEAM education in Türkiye and Europe.

The EDUSIMSTEAM project included the following:

- Learning Scenarios,
- Teacher training on STEAM through the Scientix project,
- STEAM integrated curriculum,
- STEAM education in flexible learning zones, and
- EDUSIMSTEAM project interdisciplinary practices in STEAM and innovation development workshops.

The MoNE (2022d) gathered the results and participants' feedback regarding the EDUSIMSTEAM project under four topics as detailed below.

The first was an EDUSIMSTEAM Needs Analysis Report. The participants were 1,108 teachers from Türkiye and project partners from the Netherlands, Ireland, Lithuania, Portugal, and Spain. Data were collected through surveys and interviews conducted between February and October 2020. The resultant report included the needs of teachers for STEAM education and provided pedagogical references regarding the educational policies to be implemented throughout the project. The conclusion of the report indicated that although teachers from the participating countries had some knowledge of STEAM training, they needed more training in the application of STEAM (MoNE, 2020d). This report also showed that a significant number of teachers in Türkiye were not yet sufficiently equipped to teach STEAM. The lack of preparation of these teachers to integrate arts education into STEAM is a strong indication that the inclusion of art is still incomplete in STEAM education.

Second, Learning Scenarios constituted another report prepared within the framework of the EDUSIMSTEAM project and was designed to explain the development of 22 learning scenarios, the scenario topics, and their execution (MoNE, 2022a). These learning scenarios on STEAM at the middle-school level (for students aged 10–14 years) were developed in cooperation with project partners Vilnius University, the Helix5 Company, and Middle East Technical University (METU). For example, the following learning scenarios were designed to help students develop robotics literacy and algorithmic thinking skills:

- Street Lighting in the Smart City
- Waste Collection in the Smart City
- Our Mission to Mars
- Various Themes for the Middle Level
- The COVID-19 Outbreak
- Transportation in the Smart City

I discussed this report with Prof. Erdinç Çakıroğlu from METU, one of the experts working on the project. In our meeting, Çakıroğlu said that the learning scenario simulations were the contributions of companies contracted for the project, but the pilot applications had not yet been created (personal communication, October 11, 2022). He stated that the content of the STEAM-based scenarios was under construction, while the subject of art was handled in the context of humanities and industrial design. According to Prof. Çakıroğlu the Ministry was looking for an art educator with expertise in STEAM education to work on both the individual STEAM scenario issues and the entire EDUSIMSTEAM project. This made the lack of experts in the field of art education abundantly clear. Although there was no specific information about art being included, the developed scenarios would still diversify STEAM teaching.

Third, the Teacher Training Platform was piloted with 466 participants in Türkiye between September 27 and October 25, 2021, and in Spain between November 3 and December 5, 2021. As the documents and materials related to the content of this training were inaccessible to me, I have no information on the presence of art in the STEAM training.

Last, a guide for STEAM scenarios was created to help teachers obtain answers to their questions and increase their knowledge and awareness of STEAM education. Based on the information given about the EDUSIMSTEAM project and its results, I understand that the MoNE started STEAM education in Türkiye with the EU projects and not before. As the information available on the web does not include enough examples from the practices carried out in public schools, it was not possible to understand the kind of content that was created concerning art in STEAM education. Nevertheless, I am certain that the EDUSIMSTEAM activities carried out by the MoNE held some promise for the transition from STEM to STEAM in schools. However, it will take time for STEAM education to become widespread throughout the country. The number of teachers who apply STEAM learning scenarios must increase; such teachers can then update the curriculum and advance STEAM education applications.

2.2 *Studies Featuring School-Centered Applications*

I collected and summarized information about school-centered STEAM applications in public and private schools. Generally speaking, private schools are more entrepreneurial than public schools in terms of promoting STEAM education.

2.2.1 STEAM Applications in Public Schools

Public schools have very few STEAM activities. These schools mostly participate in STEAM activities by taking part in projects carried out by the MoNE. STEAM, however, is used effectively at institutions called Science and Art Centers (BILSEMs), where gifted and talented students are trained. BILSEMs, which were established by the Ministry, are not schools per se. They were established to support the education of gifted students in their own schools and to develop their talents and creativity in the fields of science, technology, mathematics, social studies, and art. For example, the Halil İnalcık Science and Art Center offers STEAM training and has prepared a STEAM Guide Activity Booklet as part of the Much More with BILSEM STEAM project backed by the EU. The booklet explains that STEAM skills are necessary in all areas of social life and that teachers have important duties to imprint the direct relationship between STEAM and life onto the minds of the students (Halil İnalcık Science and Art Center, 2022). This Center was also preparing to send seven teachers to Portugal for STEAM training as part of the Much More with BILSEM STEAM project.

2.2.2 STEAM Applications in Private Schools

Some private schools have attempted to add STEAM to their curricula by developing programs to train their teachers to understand and teach STEAM. They have also organized STEAM events in cooperation with some nongovernmental entities; for example, Nesibe Aydın Private Schools in Ankara and the Visual Arts Education Association (GÖRSED) jointly organized a STEAM conference in 2018. For this event Dr. Yichien Cooper, an expert in STEAM education from the United States, was invited to give lectures and conduct workshops. She provided information about the history of STEAM at the conference, which was attended by art teachers and educators from universities as well as by teachers at the school. Later, Dr. Cooper invited the participants to perform STEAM applications during workshops.

The METU Development Foundation School (or ODTÜ Geliştirme Vakfı Okulu) is a private school in Ankara that has implemented STEAM education. Here, STEAM concepts are applied at the middle school level, in Grades 7 and 8 (which include students 14–15 years of age). During an interview with me, Technology and Design teacher Güneş Öncel said that students identify problems and design prototypes using the stages of engineering design (personal communication, October 28, 2022). The engineering design process, prototype, presentation, and advertising–marketing were the key elements of the innovative projects that the students prepared throughout the year. Ultimately, they took part in the Technology and Design Festival held at the end of the year, presenting their projects to a jury and an audience.

As an example, students used Tinkercad, a 3D design application described by teachers as "easy to use," to make prototypes of their designs in the Architectural Design unit. Students were also given introductory constructive solid geometry training that enabled them to create 3D-printed models. Specifically, the students made architectural designs associated with two themes: first, Creative Worlds and Architecture and second, Metaverse Architecture.

In another example of STEAM implementation, a static construction competition, students learned about balance, static subjects, and mathematical calculations. The students received a certain number of chickpeas and toothpicks and were expected to use the most appropriate number of items necessary to create the most balanced static structure. The designers whose structures carried the most weight were rewarded.

A student who created 3D prototypes explained his design as follows:

> The community for which I designed the settlement is a miners' community and a group of people who have become very rich in this way. This community sits inside the mountain, and its population is very small. The people of this community do not like to live together, and there are class differences between them. They also yearn to see the sun because they live in the mountains. (personal communication, November 4, 2022)

As shown in Figure 4.2, the architectural design and its 3D-printed model show that the miners' community appears like two separate orbs connected by a path. The red square on one of the orbs represents the door to the community. The design of the white structure and its 3D-printed model represent a dwelling inside the mountain where a small group of people live.

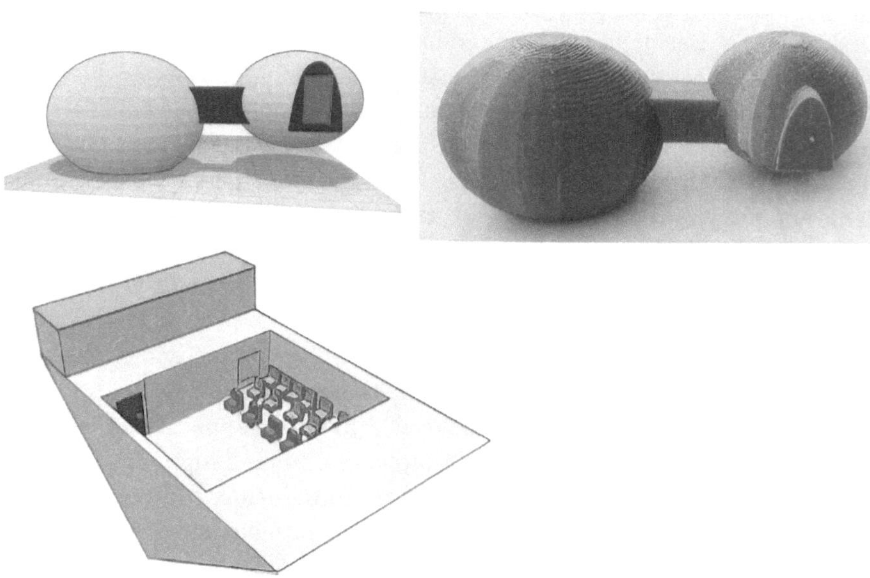

FIGURE 4.2 STEAM education applications of a Grade 7 student, 2022
PHOTOGRAPH: ODTÜ GELIŞTIRME VAKFI OKULU

As part of the Technology and Design Festival, a group of Grade 8 students set out to bring game designs to life. They designed products, such as logos, game floors, user manuals, pawns, and dice, through 3D modeling and visualization programs in their Design and Informatics workshops. The games were designed with the aim of teaching topics chosen by the students from the fields of mathematics and science. Students filmed skits to promote and advertise their designs before presenting their work to jury members from various fields, who evaluated advertisements and products and presented awards.

According to my interview with Güneş Öncel, Grade 7 students participated in the Robot Olympiad with teams they formed according to their interest in technology and design, coding, and mathematics (personal communication, October 24, 2022). During this event students competed by designing spaces and coding to solve a STEAM problem. Their creations were evaluated by jury members from different disciplines. These examples show that STEAM education taking place within the programs of METU Private School was applied deliberately and with care.

In addition, in 2022 I conducted interviews with the teachers working in the TED Private School and the SINAV Private School. From these interviews, I learned that they had received their STEAM training online from private information technology companies through the initiative of school administrations. These teachers stated that in their STEAM training they focused mainly on lesson applications related to the integration of science fields, including physics, chemistry, and biology. They tried to implement the STEAM education in the form of project activities within the clubs at school rather than during classroom lessons.

2.3 *Studies Carried Out at Universities*

STEAM education studies carried out at universities in Türkiye can be classified as those aimed at enhancing their own education programs, those conducted by researchers, and those conducted for public and private schools. Research conducted at universities with colleges of education that prepare future teachers is generally focused on the development and educational outcomes of the primary and secondary education curricula. National and international publications produced by the university faculty members focus mostly on STEM, rather than STEAM. A few articles on applications have emphasized art in STEAM studies (Başaran & Erol 2021; Erdoğan, 2020; Kahya & Özdilek 2021; Mercin, 2019).

STEAM activities are seldom carried out in these colleges of education. In a rare case the METU STEAM Day event was held in 2019. It was organized in cooperation with the Middle East Technical University BİLTEMM Center and the METU STEAM Student Group. This event included talks such as "Design Factory: Interdisciplinary Design Studio" and "From STEM to STEAM Education: Theoretical Framework and Historical Development," but most of the talks still focused on STEM content (Middle East Technical University [METU], 2019b).

Some colleges added STEAM projects to improve their own curricula; for example, the METU College of Architecture implemented a STEAM project called ArchiSTEAM with architecture schools in Aalborg Universitet in Denmark and the University of Bologna in Italy (METU, 2019a). As part of the project, the College of Architecture created lessons incorporating computer design programs into the curriculum. In another example, the Marmara University Continuing Education Center carried out a paid activity called the BITEKMUSEM (STEAM) Education Program in 2019 to introduce STEAM to primary and secondary school teachers and students and to exhibit the applications of STEAM. This program covered robot construction workshops and explained analytical geometry with origami art as well as the role of mathematics in music and basic music education.

3 Concluding Remarks

Although the STEAM studies and activities described above were valuable, I believe that the desired level of implementation of STEAM education has not been reached in Türkiye. This is because STEAM education is not yet fully understood. STEAM education has three components: creative design, an emotional element, and integration of content (Park et al., 2016). Creative design refers to a comprehensive process in which students use their imagination and aesthetic sense to find optimal solutions to problems. The emotional element enables positively oriented learning cycles, wherein students develop feelings of interest, confidence, intellectual satisfaction, and achievement as they discover motivation, passion, flow, and personal meaning through learning. The integration of content aims to connect all disciplines to real life from a holistic perspective. Such knowledge along with experimentation of STEAM education has not yet been significantly shared and discussed among teachers and educators in Türkiye.

Considering the few STEAM applications carried out in schools in Türkiye, it can be said that educators still do not fully understand the difference between STEM and STEAM. They do not comprehend which fields of activity are covered by the art component in STEAM. This was determined in my interviews with private school teachers who had received STEM or STEAM education. Most of these teachers had received their STEAM training through online training programs conducted by companies school administrators had hired. When the content of these programs was examined, science was shown to be dominant, while art was limited to 3D design programs. Thus, education is still STEM-oriented. I speculate that the most important reason for this is insufficient numbers of educators and experts to integrate the field of art into STEAM.

Despite the limited understanding and integration of the A in STEAM in Türkiye, today the interest in STEAM education is high. Private schools especially use STEAM, not only for educational purposes but also as an important element in marketing for their schools. There are three main reasons for the high attention in STEAM education:

- Türkiye's integration into the European Union education system,
- reports prepared by employer organizations that emphasize the importance and necessity of STEAM education and recommend it to the government, and
- existing academic research on the importance of STEAM education.

The Organization for Economic Cooperation and Development prepared a report containing 21st century skills and competencies and recommended it to member countries (Ananiadou & Claro, 2009). After examining this report, I can say that STEAM training can play an important role in developing those skills. STEAM is an interdisciplinary integrated education model that can clearly contribute to the knowledge base of an individual equipped with 21st-century skills. I believe the STEAM approach will continue to develop globally in the coming years. In other words the discipline of art will always play a role in education, and it will make a significant contribution to the growth of the individual in harmony with other disciplines.

References

Ananiadou, K., & Claro, M. (2009). 21st century skills and competences for new millennium learners in OECD countries. *OECD Education Working Papers*, 41. OECD Publishing. https://doi.org/10.1787/218525261154

Avcı, F. (2021). Çevrim içi bir öğrenme ortamı olarak eTwinning platformuna ilişkin öğretmenlerin görüş ve değerlendirmeleri [Teachers' opinions and evaluations of the eTwinning platform as an online learning environment]. *Cumhuriyet International Journal of Education*, *10*(1), 1–22. http://dx.doi.org/10.30703/cije.663472

Başaran, M., & Erol, M. (2021). Recognizing aesthetics in nature with STEM and STEAM education. *Research in Science & Technological Education*. https://doi.org/10.1080/02635143.2021.1908248

Başaran, M., Kaya, Z., Akbaş, N., & Yalçın, N. (2020). Proje tabanlı öğretim sürecinde eTwinning faaliyeti'nin öğretmenlerin mesleki gelişimlerine yansıması [Teachers of eTwinning activities in project-based teaching process reflect on their professional development]. *Eğitim Kuram ve Uygulama Araştırmaları Dergisi, 6*(3), 373–392.

Çiftçi, A., Topcu, M. S., & Erdoğan, İ. (2020). Gender gap and career choices in STEM education: Turkey sample. *International Journal of Progressive Education, 16*(3), 53–66. https://doi.org/10.29329/ijpe.2020.248.4

Durak, G., Çankaya, S., Nacak, A. F., & Baysal, F. E. (2021). The current state of Turkish STEM research: A systematic review study. *Necatibey Faculty of Education Electronic Journal of Science and Mathematics Education, 15*(2), 383–403. https://doi.org/10.17522/balikesirnef.1032295

Erdoğan, S. (2020). STEAM ve sanat eğitimi ilişkisi [The relationship between STEAM and art education]. *Selçuk* Ün. *Sosyal Bilimler Enstitüsü Dergisi, 2020*(44), 303–316.

European Union. (2020). *The STE(A)M partnerships programme of Scientix*. http://www.scientix.eu/projects/steam-partnerships

European Union. (2021). *European Innovation Platform for Cultural and Creative Industries*. https://eip21.eu/

European Union. (2022). *e-Twinning Project*. https://www.etwinning.net/tr/pub/index.htm

Gündüz Çetin, İ., & Gündoğdu, K. (2022). eTwinning proje uygulamalarına ilişkin öğretmen görüşleri [Teacher opinions on eTwinning project implementation]. *Elektronik Sosyal Bilimler Dergisi, 21*(81), 76–90. https://doi.org/10.17755/esosder.981142

Halil İnalcık Science and Art Center. (2022). *Halil* İnalcık *Bilim ve Sanat Merkezi STEAM projeleri* [*Halil* İnalcık *Science and Art Center's STEAM projects*]. Güncelleme [Update]. https://nilhibilsem.meb.k12.tr/icerikler/bilsem-steam-ile-cok-daha-fazlasi_12403886.html

Kahya, V., & Özdilek, Z. (2021). Alan uzmanlarının STEAM eğitimi ile ilgili görüşleri [Opinions of field experts about STEAM education]. *Atatürk* Üniversitesi *Kazım Karabekir Eğitim Fakültesi Dergisi*, (42), 122–147. https://doi.org/10.33418/ataunikkefd.862411

Mercin, L. (2019). STEAM eğitiminde sanatın yeri [The place of art in STEAM Education]. İnönü Üniversitesi *Sanat ve Tasarım Dergisi, Haziran, 9*(19), 28–41. https://doi.org/10.16950/iujad.514132

Ministry of National Education. (2016). *STEM eğitimi raporu* [*Report of STEM education*]. http://yegitek.meb.gov.tr/STEM_Egitimi_Raporu.pdf

Middle East Technical University. (2019a). *ARCHISTEAM: Greening the skills of architecture students via STEAM education.* https://archisteam.metu.edu.tr/

Middle East Technical University. (2019b). *Middle East Technical University BİLTEMM Center.* http://biltemm.metu.edu.tr/en/

Ministry of National Education. (2022a). *EDUSIMSTEAM learning scenarios.* https://edusimsteam.eba.gov.tr/wp-content/uploads/2022/06/EDUSIMSTEAM-Ogrenme-Senaryolari-v2.pdf

Ministry of National Education. (2022b). *EDUSIMSTEAM okullarda STEAM eğitimin yaygınlaştırılması* [*Fostering STEAM education in schools*]. https://edusimsteam.eba.gov.tr/?page_id=95&lang=en

Ministry of National Education. (2022c). *eTwinning.* http://etwinning.meb.gov.tr/

Ministry of National Education. (2022d). *Outputs of the Fostering STEAM Education in Schools project.* https://edusimsteam.eba.gov.tr/?page_id=1135

Park, H., Byun, S., Sim, J., Han, H., & Baek, Y. S. (2016). Teachers' perceptions and practices of STEAM education in South Korea. *Eurasia Journal of Mathematics, Science & Technology Education, 12*(7), 1739–1753. https://www.ejmste.com/download/teachers-perceptions-and-practices-of-steam-education-in-south-korea-4573.pdf

Subramaniam, B. (2021). Why adding the arts to science education makes us better-rounded. *The Hindu.* https://www.thehindu.com/education/why-adding-the-arts-to-science-education-makes-us-better-rounded/article37931481.ece

Taylor, P. C. (2017). Enriching STEM with the arts to better prepare 21st century citizens. *AIP Conference Proceedings, 1923*(1). https://doi.org/10.1063/1.5019491

Yılmaz, A., Gülgün, C., Murat Çetinkaya, M., Doğanay, K. (2018). Initiatives and new trends toward STEM education in Turkey. *Journal of Education and Training Studies 6*(11a), 1–10. https://doi.org/10.11114/jets.v6i11a.3795

PART 2

Redirecting STEAM Foci in the Communities

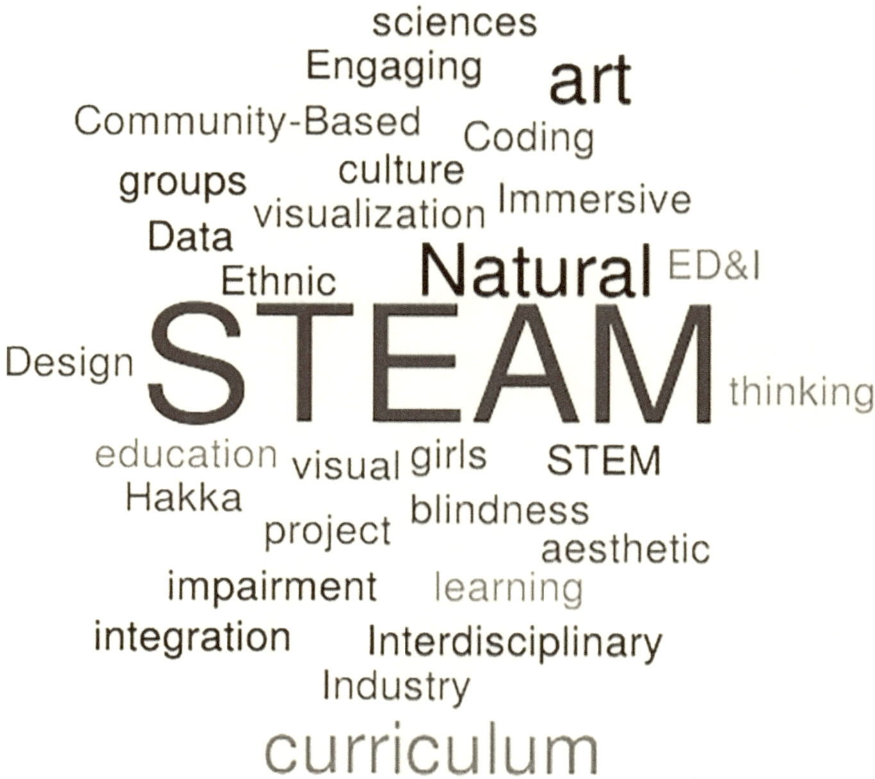
sciences
Engaging
art
Community-Based
Coding
groups
culture
Immersive
visualization
Data
Natural
ED&I
Ethnic
Design
STEAM
thinking
education
visual
girls
STEM
Hakka
blindness
project
aesthetic
impairment
learning
integration
Interdisciplinary
Industry
curriculum

CHAPTER 5

Reflections on the Practice of STEAM Education in Penang, Malaysia

Paul Boey

Abstract

This chapter focuses primarily on the need for and practice of STEAM education in the school systems of Penang State, Malaysia. Recognizing the challenges in developing and promoting STEAM education, the author affirms the value of school-based and community-based STEAM projects. Notably, the impact of STEAM education on students includes enhancing their creativity, problem-solving abilities, teamwork skills, and leadership qualities. Based on the author's observation of and leadership in STEAM initiatives, he perceives positive responses from teachers and students. Looking ahead, the author offers a critique of STEAM education in Penang and suggests that the Penang State school systems bolster teacher training and resource allocation to ensure high-quality STEAM education for all students.

Keywords

STEM to STEAM – coding – industry 4.0 – community-based STEAM project

1 Introduction

In recent years the emphasis on science, technology, engineering, and mathematics (STEM) education in Penang, Malaysia, has grown stronger. With the arrival of Malaysian Industry 4.0, STEM education has emerged as a critical area of focus among the industrial and educational sectors in the country (Government Policies, 2019). Malaysian government and private educational institutions have expanded their investments in equipment and resources, aiming to attract students to technological and robotics programs. As the value of "A" (i.e., Art) in strengthening STEM learning has gained traction, schools have initiated the revision of the STEM curriculum to include design thinking and hands-on creative projects or the development of new STEAM initiatives.

 | DOI:10.1163/9789004714748_006

Thus, one may safely assume that the education system in Malaysia is gradually evolving; nevertheless, while Penang stands as a leading STEAM hub in Malaysia, its STEAM curriculum notably remains somewhat STEM-centric, indicating the need for further efforts in integrating the arts with various STEM subjects. The endeavor of transforming STEM to STEAM, or STEM and STEAM, is vital to ensure that students receive a comprehensive education and are well prepared to meet the future demands of the workforce and society (Wan et al., 2019).

2 The Era of Malaysia's Industry 4.0

The Malaysian National Industry 4.0 Policy, referred to as Industry 4.0, was launched on October 31, 2018, by former Prime Minister Mahathir. This initiative was a call to action by the Malaysian government, urging manufacturing and related service companies to undergo digital transformation. The primary goal was to encourage these companies to achieve more systematic and comprehensive advancements in various areas, including human resources, manufacturing processes, and technology. This approach was designed to foster the growth of smarter and more robust enterprises of the manufacture of five selected products: electrical and electronic products, mechanical equipment, chemical products, aerospace technologies, and medical equipment. As for the service industry, it emphasizes key sectors, such as primary hubs, logistics, ecosystems related to e-commerce, green technology, renewable energy, and waste management (Government Policies, 2019).

In response to the Malaysian government's pursuit of the Industry 4.0 policy, aimed at achieving advanced urbanization and industrialization management, the Penang state government and various universities have recognized the significance of nurturing students' scientific and technological talents. The Penang State government has allocated approximately RM60.5 million (equivalent to around 13.56 million USD) for the establishment of several STEM education centers, including the Penang Science Cluster, Tech Dome Penang, Karpal Singh Penang Learning Center, aCAT Penang, and Penang Digital Library (The Launching of Penang STEM SDN BHD, 2017).

In addition, the state has been a host to significant events in the field, such as the 8th International Conference on Mathematics and Science Education and an Innovation and Technology Exhibition, both in 2019. These events attracted researchers and educators from more than 10 countries, fostering collaboration and the exchange of knowledge.

Based on the Malaysia Education Blueprint 2013–2025, STEM education has been integrated into the K–12 curriculum (Ministry of Education Malaysia, 2013). Many K–12 schools offer STEM extracurricular activities, such as the International Math Olympiad and robotics courses. In higher education, for example, Taylor's University in Subang Jaya, a forward-looking approach has merged engineering and computer science courses. This fusion approach provides graduates with enhanced learning opportunities for the fast-paced and technologically driven career market of the 21st century. In response to the demands of this evolving market, the universities offer degree programs specifically dedicated to robotics design.

3 Challenges in Sustaining STEM Education

With the rapid growth of manufacturing industries in Malaysia, a demand arose for future talents who could creatively and effectively incorporate STEM knowledge. Managers of both local and international factories are eager to hire technology professionals. The urgency of the need is particularly evident in Penang, a city dominated by industrialization and technology. In response to the call, the education sector in Penang has emphasized the development of STEM, with a focus on integrating creative arts as an additional goal. Although the intention is strong, this transformation has not been without its challenges.

According to Academy of Sciences Malaysia (ASM) (2017), the Malaysian government still maintains the 60:40 policy formulated in 1967, whereby the ratio of science students to liberal arts students is 60:40; however, this ratio appears to be difficult to achieve. ASM reported that STEM enrollment among Secondary Form 4 and Form 5 students aged 16–17 dropped from 48.15% in 2012 to 45.74% in 2017. Comparing the data from the Ministry of Education, Idris et al. (2023) summarized that STEM enrollment after a brief increase to 47.18% in 2020 had dropped to 40.95% in 2021. Moreover, while the percentage of students in STEM was 47.18%, only 20.51% involved in pure science subjects; the remaining 26.67% received technical training in vocational education. In general, a decline in STEM enrollment has occurred in recent decades.

I speculate several reasons leading to a downturn in STEM student populations in Malaysia. First, teachers and parents still emphasize learners' academic achievements, often the result of passive teaching and learning methods. To many students, the content of STEM courses is too difficult to comprehend. The rote and rigid content delivery fails to arouse their interest. Second, concerned with their overall GPA, many students strategically avoid taking STEM courses.

Third, students are puzzled by the usefulness of STEM knowledge in everyday life and how well STEM knowledge can prepare them for the real world.

From the viewpoint of teachers, a significant obstacle for them to shift from a STEM to a STEAM approach involves the undoing of the existing classification of STEM as science and STEAM as art in education administration departments. This inflexible categorization creates challenges for teachers interested in implementing a comprehensive STEAM curriculum across multiple disciplines. For instance, art teachers may have difficulty accessing the necessary hardware and equipment typically located in science or computer rooms when preparing to integrate art into science and technology lessons. This equipment gap further hinders their ability to facilitate students' needs when learning STEAM subjects. These issues undoubtedly pose significant hurdles in the process of developing creative STEAM school projects. Thus, examining the existing educational practices and structures is essential in better supporting the integration of the creative arts with STEM disciplines.

4 The Transformation of STEM and STEAM Education in Penang

As an alternative solution to cultivating more creative talent without constraint by existing educational systems, more and more so-called STEAM learning centers have popped up in various high schools. For example, Penang Han Chiang High School launched the STEAM Lab—STEM + Art to enhance its STEAM education courses. Similarly, in 2014, Penang Chung Ling Independent High School added the Creative Electronics Club to its extracurricular activities to inspire students' understanding of electronics technology. These STEAM learning spaces are designed to build a more creative learning atmosphere to stimulate students' interest in and potential to learn about science, technology, engineering, mathematics, and art.

In essence, I believe that as schools offer more STEM courses, educators will realize the benefits of the inclusion of the creative arts in these programs. I am confident that in Penang current technology-heavy talent training programs will be transformed to STEAM education through continuous efforts in searching for more suitable and adaptable learning methods. I also believe that more mentor resources will be available to facilitate educators in developing comprehensive learning and experiential STEAM curricula. In what follows I introduce three cases exemplifying the transformation of STEM to STEAM, or STEM and STEAM.

4.1 *STE(A)M Education in Penang Han Chiang High School*

In 2017 Penang Han Chiang High School (PHCHS) introduced STEAM education courses into its curriculum. By integrating technology education into the high school program, PHCHS provided students with the opportunity to learn mechanical design and software editing and engage in various local and international competitions, aiming to secure awards. Yeap Jin Bo, the STEAM course director at PHCHS, promotes the five cores of STEAM literacy, namely, transdisciplinary learning, hands-on experience, real-life applications, problem-solving, and multisensory engagement (*How to Create*, 2020). His approach to STEAM education has garnered support from the school, parents, and the broader society.

Meanwhile, PHCHS has launched a digital library and an innovative technology lab, where students are encouraged to design and research technology products and master a variety of fundamental skillsets in digital technology, including computer program development, 3D printing, and robot building. At PHCHS programming in technology is viewed as a creative agent, taking practical and transdisciplinary approaches in STEAM education. The goal is to foster hands-on experimentation and active inquiry among students, cultivating their ability to solve problems creatively.

In the STEAM Lab—STEM + Art, the school uses Arduino as a mechanical platform and software editor, allowing students to acquire hands-on experience in STEAM and develop their understanding of STEAM content. This innovative technology lab fosters cross-disciplinary and creative thinking in students, departing from traditional, single-discipline learning methods and enhancing students' problem-solving abilities.

Mr. Yeap stated that STEAM education has gained popularity because it has nurtured scientific and technological talents while incorporating the art and humanities as the foundation of STEAM (*How to Create*, 2020). He believes that education in the arts and the humanities helps enhance students' critical thinking as they transform their ideas into tangible objects; moreover, following the practice in the art and humanities, students collaborate in teams and engage in discussions. Consequently, they respond to questions with more than rote answers, cultivating their ability to flexibly apply their knowledge to investigate and solve STEAM problems. This kind of STEAM education can further equip students with multiple approaches to address real-life problems, making STEAM a valuable and applicable tool in both science and everyday life.

4.2 *STE(A)M Education in Penang Chung Ling Independent High School*

Penang Chung Ling Independent High School (PCLIHS) has also promoted the transformation of STEM to STEAM education. The goal at PCLIHS is to enhance students' creativity through a student-led initiative—the Creative Electronics Club—which has taken a site-specific and practical approach, actively contributing to the improvement of school safety and the upgrading of some outdated equipment through their STEAM projects.

For instance, during the December 2017 Penang floods, PCLIHS suffered significant losses in equipment, leading to life-threatening situations in the dormitories with students trapped in the rising water. Afterwards, PCLIHS initiated discussions on how to predict dangerous water levels in advance to notify dormitory students to evacuate or seek safety. These discussions inspired students from the Creative Electronics Club to create flood alarms. Leveraging their long-term training in STEAM education with a focus on design thinking, the students provided a simple solution: using a pair of iron chopsticks as a contact pin to conduct electricity and developing a flood disaster siren (*How to Create*, 2020).

The students created the flood alarm by applying their knowledge of scientific theories and their hands-on programming, electrical, and engineering skills. Here is a synopsis of their idea: When the water level reaches the alarm stylus, the alarm bell sounds, and a warning light flashes, triggering the Global System for Mobile Communications, which simultaneously sends text messages to the principal and teachers. This prompts immediate emergency arrangements to ensure the safety of everyone. I am impressed by the students' innovative and practical solution. From brainstorming ideas to building a prototype, they accomplished this in just a month and a half with a budget of RM300 (approximately USD $68).

Another example involves the concern at PCLIHS about the frequent malfunctions of the machinery in the ancient bell tower and the mounting repair costs associated with its mechanical problems. This real-life problem was presented to the students in the Creative Electronics Club, who were tasked with replacing the traditional bell tower with electronic alternatives. Guided by experts in electrical engineering, students successfully fulfilled their goal. This example further illustrates the commitment at PCLIHS to using STEAM education to develop skills that are highly adaptable in real-world society. It also offers students an opportunity to see how relevant and applicable both STEM and design thinking (A) are (*How to Create*, 2020). PCLIHS is dedicated to this direction, hiring experts and teachers to provide students with systematic educational programs, motivating them to acquire basic knowledge in computer hardware and computer software operating systems. Students not only learn but also master drill-and-practice software programming, gradually progressing to real-world programming scenarios that require complex thinking and practical application.

4.3 *Community-Based STEAM Education*

To many teachers and students, the concept of STEAM is foreign. As a supporter of community art, I believe engaging students with critical issues in the community through arts is an effective way to enrich STEAM education. For example, I carried out a STEAM-related environmental awareness project called Recycle +. The process of assembling scrap tires to create a large scale of sculpture requires an artistic vision in addition to math, engineering, and physics. This outdoor installation featured a dusky leaf monkey, an endangered species in Myanmar, Malaysia. The use of discarded tires makes a sad but profound statement about how rapid urbanization has threatened dusky leaf monkey's habitat. Through this STEAM project, I raised awareness about the harmful effects of hunting and deforestation on wildlife (Figure 5.1).

To echo my large-scale STEAM project and to continue fostering the development of community art education, I collaborated with local and International Baccalaureate schools to execute a series of exhibitions on animal protection. Following the successful art education initiative in Penang, known as ART for Wildlife Conservation, in 2019 I embarked on a STEAM project in Muar, Malaysia, involving the integration of science and community art. I took the lead in guiding teachers and students to create an installation art piece depicting a stranded whale. This remarkable artwork was placed at Pu Nan Primary School as part of the *Heart of the Ocean* community art education exhibition.

FIGURE 5.1 I used scrap tires to create an installation art, depicting an endangered dusky leaf monkey in Myanmar, Malaysia

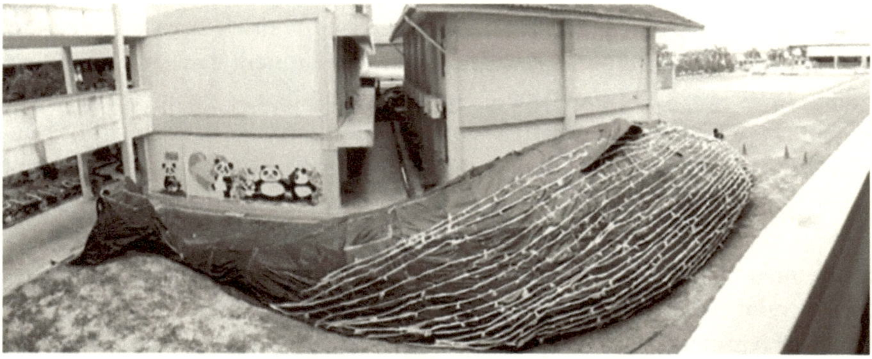

FIGURE 5.2 For the *Heart of the Ocean* community art education exhibition, I collaborated with teachers and students to create an installation piece on campus depicting a stranded whale. The purpose of this collaborative effort was to inspire others and raise awareness about important topics related to environmental protection

In an effort to promote marine science and environmental awareness, *the Stranded Whale* was constructed with recycled materials; specifically, the participating students used plastic bags and applied bamboo weaving techniques to create a life-sized whale sculpture. During the process of making the whale, students not only developed a deeper understanding of proportion but also gained new insights into the anatomy and bodily structure of a whale.

In order to convey the detrimental impact of ocean waste on marine animals, the students decided to position their whale in a "stranded" posture. This powerful visual representation served as a poignant reminder of the challenges faced by marine life due to human beings' careless waste dumping and environmental pollution (Figure 5.2).

5 Reconsidering Traditional Art Courses: Let STEAM Education Pave the Way for Youth in the Future

Although Penang's STEAM education may not have progressed as rapidly as that in some other countries, in the current era of Industry 4.0, schools in Penang must cultivate talents in scientific and technological innovation for the 21st century. To begin this quest, I believe Herbert Read's (2007) *Education Through Art* could play a pivotal role in supporting government efforts in Malaysia to promote STEAM education because the inclusion of arts disciplines like art, music, and design can significantly contribute to the development of students' overall skills, ultimately leading to higher-level performance in STEM subjects.

The "A" in STEAM, can encompass various aspects, including spontaneous art creation, art appreciation, and personal expression. Arts education emphasizes the process of critical thinking, creativity, design thinking, and innovation based on comprehensive understanding of the subject in question. In other words, incorporating art, music, and design elements provides students with opportunities to exercise their diverse intellectual capacities, fostering autonomous and creative problem-solving skills. This, in turn, ensures the cultivation of well-rounded talents in STEAM and strengths, such as IQ and EQ. This can certainly nurture students' extensive abilities, such as imagination, creativity, cognition, perception, and expression.

As children undergo distinct physical, mental, and emotional development at varying ages, the STEAM curriculum can offer suitable learning content to align with their developmental characteristics. The incorporation of art in the U.S. concept of STEAM education can be adaptively transformed through the use of technology. This adaptation allows for the provision of a learning platform that integrates hands-on activities and experiences, taking into account the unique physical and mental development stages of each child's growth. Such an approach not only addresses their individual developmental needs but also effectively fulfills educational objectives that encompass strengthening critical thinking, communication skills, teamwork, creativity, and innovation.

At the time of this writing, a significant number of art teachers in Malaysia continued to adhere to conventional art teaching methods outlined in government-issued art textbooks. This traditionalist approach to art education can be exemplified through the art curricula offered in two key categories of high schools in Malaysia: government-funded Chinese high schools and private or independent Chinese high schools.

Government-funded schools typically structure their art curricula around subjects like fine art, visual communication arts, design, visual art history, and traditional handicrafts among others. The content of these curricula places a strong emphasis on fundamental elements of art, such as lines, shapes, colors, and more. It also covers core design principles of art, including concepts like contrast, balance, and unity. In addition, students explore and use conventional and traditional art mediums to design and craft as part of the coursework. The assessment of student learning is based on proficiency in art-related technology (media art), art technique, art knowledge, and their ability to reflect Malaysia's historical and cultural aspects as well as its artistic traditions through their artworks.

In contrast, private or independent schools provide a more extensive and diverse array of art learning content, which encourages students to delve into the philosophical aspects of art, exploring topics like the meaning of art, aesthetics, and the intersection of art and life. Despite this broader approach, the learning outcomes in these schools still prioritize student mastery in art-related technology, art studio technique, and fluency in specific art content knowledge. In essence, both types of schools share a common focus on teaching foundational principles of art and studio techniques.

Art educators practicing a traditional art education approach face challenges when it comes to embracing the STEAM curriculum for integrating science and art. Currently, the significance of STEAM education is not recognized in every school in Malaysia. As a result, I believe the government's capacity to promote and advocate for educational reforms is limited. In my observation, one of the key barriers preventing art students from benefiting from the 21st-century STEAM approach is that many educators are not fully aware of the creative era in which they teach and live. Consequently, they have not implemented the necessary changes to equip their art students with the skills and perspectives demanded by the dynamic contemporary creative industry.

Some art teachers, who adhere to tradition, may question the extent to which art courses can prepare children for the changes and challenges brought about by the rapid economic, scientific, and technological advancements in society and the world. Their seemingly defeatist mindset and attitude may hinder their willingness to adopt STEAM teaching methods.

I believe that the incorporation of art into the STEAM curriculum can ignite children's interest in science, technology, engineering, mathematics, and other subjects while broadening their horizons on the sea of knowledge. The inclusion of art subjects opens up new perspectives for implementing STEAM teaching methods that may not have received adequate attention in the past. For instance, we can guide students to seamlessly integrate the knowledge they have acquired in each subject through art. Artistic learning not only enables them to connect science and art to their daily lives but also enriches the content of art education. Art-based learning nurtures the ability to transform abstract concepts into tangible objects, fosters imagination, promotes creative thinking, and enhances expressive skills. As a result, art teachers can effectively harness the power of art to gradually and profoundly help children grasp the learning objectives of STEAM, creating an environment of active and engaged learning.

6 Final Reflection

In the context of the Malaysian Industry 4.0 revolution, an unprecedented need exists to place a strong emphasis on fostering innovation, analytical skills, strategic thinking, and critical reasoning. The school projects described in this chapter serve as compelling evidence of how STEM knowledge can be significantly enhanced through STEAM curriculum emphasizing hands-on creation, artistic practice, and the incorporation of design elements. In this way, STEAM education can fulfill the need in Malaysia; however, the school projects I exemplify primarily existed in selected independent Chinese schools. I attribute such limitation to a misconception that implementing a STEAM curriculum is costly because it requires a substantial allocation of resources, equipment, and teachers. Other school officials interested in implementing STEAM curriculum are overconcerned with inadequate funding and become hesitant to establish or expand their STEAM curriculum to provide creative learning opportunities in support of students' learning.

To continue advocating for STEAM education, I propose a restructuring of school systems, involving academic structure, resources, equipment, and teachers. Avoiding the confinement of art and design or fine art within an art department, curriculum designers should instead cross-list these classes in a range of disciplines. Such classes should be cotaught by teachers from various subject areas. Reform like this will allow STEAM education to center on the integration and interdisciplinary aspects of multiple subjects with the goal of nurturing students' practical skills, hands-on ability, critical thinking, creative problem-solving approaches, and the execution of projects involving a wide range of materials. Thus, students are better equipped to achieve breakthroughs in all facets of their learning journey; furthermore, they are provided with assorted tools and methods to explore innovative solutions, present data, and establish connections across various fields creatively.

Concerning the educational vision for STEAM education, I anticipate the Malaysian government, universities, schools, and other educational institutions to collaborate in establishing a team of highly skilled art and general teachers. The government should also provide professional development opportunities for educators to become proficient in STEAM education. This endeavor could involve the development of a national art education curriculum with a strong emphasis on addressing theoretical and practical challenges in STEAM instruction across various disciplines, both in public and private schools. This includes formulating clear STEAM education plans, tailored to various age groups and subject areas, ensuring that all students have access to STEAM education; moreover, maintaining ongoing collaboration with the creative industry is essential to help students gain practical experience. Engaging students in community-based art projects will also further their understanding of how STEAM can be applied in real life. These joint efforts can significantly contribute to the advancement of STEAM education, ultimately securing a brighter future for young learners.

References

Academy of Sciences Malaysia. (2017). Science outlook 2017: Unlocking the future. https://www.akademisains.gov.my/asm-publication/science-outlook-report-2017/

Government Policies. (2019, July 12). National policy on Industry 4.0 – Industry4WRD. *Prime minister's office of Malaysia official website*. https://www.pmo.gov.my/2019/07/national-policy-on-industry-4-0-industry4wrd/

How to create STEAM education in life situations. [如何在生活情境中创造*STEAM*教育？]. (2020, July 15). https://dzblueprint.dongzong.my/zxs-about-2/120-2020-007

Idris, R., Govindasamy, P., & Nachiappan, S. (2023). Challenge and obstacles of STEM education in Malaysia. *International Journal of Academic Research in Business and Social Sciences, 13*(4), 820–828.

Ministry of Education Malaysia. (2013). *Malaysia Education Blueprint 2013–2025*. https://www.pmo.gov.my/wp-content/uploads/2019/07/Malaysia-Education-Blueprint-2013-2025.pdf

Read, H. (2007). *Education through art*. Routledge.

The Launching of Penang STEM SDN BHD. (2017, August 19). https://cm.penang.gov.my/index.php/ms/all-stories/2185-the-launching-of-penang-stem-sdn-bhd

Wan, C. D., Sirat, M., and Razak, D. A. (2018). Education in Malaysia towards a developed nation. *Economic Working Paper, 4*, 1–20. https://www.iseas.edu.sg/wp-content/uploads/pdfs/ISEASEWP2018-4Wan.pdf

CHAPTER 6

Natural Aesthetics

Interdisciplinary Learning of the Arts and Natural Science

Chia-Hung Kao

Abstract

Utilizing the resources in the National Museum of Natural Science in Taiwan, I developed a natural aesthetics curriculum to enrich high school students' learning of the arts and natural science. Specifically in this chapter, I exemplify the three STEAM learning modules, titled Natural Science Inquiry, Interdisciplinary Integration of Art and Natural Sciences, and Friendly Environment and Sustainable Ecology. I explain the pedagogical content of the three modules and use students' work as examples to illustrate their learning of natural science, art, and selected aspects of Taiwanese culture. Emphasizing interdisciplinary and collaborative learning, students were able to acquire STEAM knowledge and enjoy practicing STEAM skills in their projects.

Keywords

natural aesthetic curriculum – interdisciplinary integration of art and natural sciences – The National Museum of Natural Science in Taiwan

1 Introduction

The National Museum of Natural Science (NMNS) in Taichung City is one of the top centers for science education in Taiwan. I received an invitation from Asia–Pacific Office for Aesthetic Education (APOAE) in 2021 to collaborate with the senior high schools to create lessons involving the resources of the NMNS. The schools participating in my natural aesthetics curriculum included Taichung Municipal Wen-Hua Senior High School, Taichung Second Senior High School, and National Changhua Girls' Senior High School. I collaborated with teachers from these schools to develop STEAM lessons that emphasize the integrated concepts of STEAM education, natural science, aesthetic literacy, and community-based education.

 | DOI:10.1163/9789004714748_007

FIGURE 6.1
Emblem of the National Museum of Natural Science

To develop STEAM lessons for the natural aesthetic curriculum, I initially adopted the core aspects of aesthetic education in Taiwan, which reflect Dasein or the existential aesthetic, the citizenship aesthetic, and the ecological aesthetic. Thus, aesthetics education emphasizes the positioning and intersection of people-to-self, people-to-society, and people-to-nature (Hung, 2012; Tsai & Liu, 2020). Second, I used the resources and embraced the values promoted by the NMNS. The resources were housed in the five main galleries in the NMNS: the Life Science Hall, the Human Cultures Hall, the Global Environment Hall, the Theaters and Science Center, and the Botanical Garden. I noticed that the design of the NMNS emblem reflects the collections in the museum as well as the value of science and aesthetics education in Taiwan. As shown in Figure 6.1, the emblem contains meaningful symbols. Below the plum blossom, Taiwan's national flower, is the letter S with its mirror image. The letter S reminds me of a Chinese jade beast headdress, a mathematical calculus, sprouting seeds, and the first letter of "science." The background colors, blue and green, in Taiwanese culture, are associated with organic live forms in natural habitats; hence, in my opinion, the design of the emblem represents the mission of the NMNS in nurturing science and the humanities.

I created a series of STEAM lessons and organized them into three main learning modules, titled Natural Science Inquiry, Interdisciplinary Integration of Art and Natural Sciences, and Friendly Environment and Sustainable Ecology, including interdisciplinary learning activities that combine studies of natural science, the arts, and the humanities. Students were also expected to complete their learning activities collaboratively. For example, centering on plants in a module, students carried out the following tasks:

- Studying information gained from a variety of educational apps (e.g., iCoBo, i-NM, and Google Classroom) and communication platforms (e.g., Canva, Slido, Padlet, and Miro)
- Working with teachers, researchers, parents, and peers to apply the information to the individual and group STEAM projects
- Building a concept-based learning atmosphere by following the APOAE manual
- Applying knowledge gained from the observation of plants in their homes, schools, and communities to discern plants in various regions in Taiwan

2 Pedagogical Content and Strategies for the Three Modules

The interdisciplinary learning approach was infused into the three modules for high school students to explore natural science and engage in aesthetics activities at the NMNS. As shown in Figure 6.2, to provide students with coherent and progressive learning opportunities, I expanded the three modules:

- Natural Science Inquiry
- Interdisciplinary Integration of Art and Natural Sciences
- Friendly Environment and Sustainable Ecology

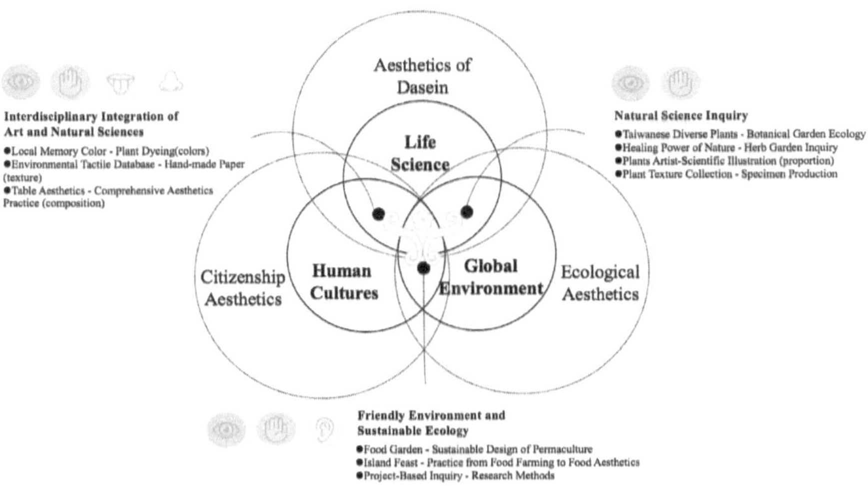

FIGURE 6.2 A conceptual map of the pedagogical content for the natural aesthetics curriculum that aligns with the three core aspects of aesthetics education in Taiwan

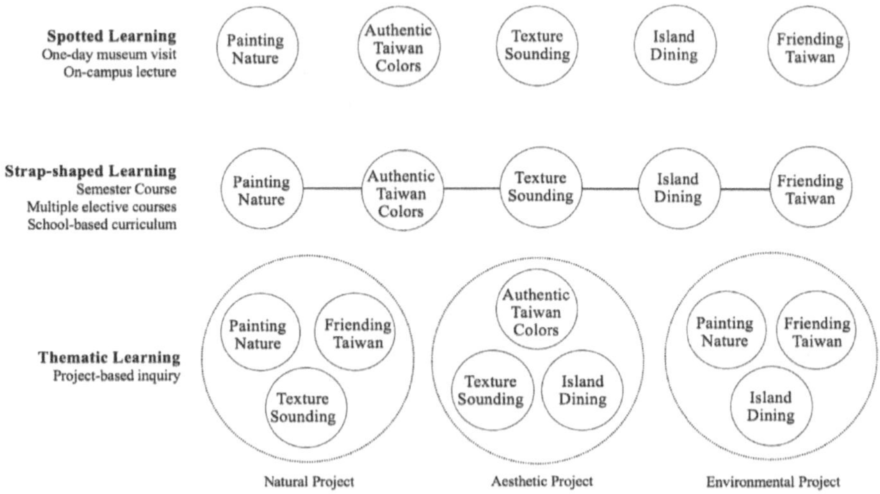

FIGURE 6.3 The spotted, strap-shaped, and thematic learning approaches

into five learning units (see Figure 6.3)
- Painting Nature
- The Sound of Texture
- Authentic Taiwan Colors
- Island Dining
- Friending Taiwan

with 10 learning activities
- Diverse Plants in Taiwan: Learning Ecology at the Botanical Garden
- The Healing Power of Nature: Inquiry Learning at the Herb Garden
- A Collection of Textures in Plants: Specimen Production
- Plant Artist: Scientific Illustration, focusing on the artistic principle of proportion
- Environmental Tactile Database: Handmade Paper, focusing on the artistic principle of texture
- Preserving Local Memory Color: Plants as Natural Dyes, focusing on the artistic principle of color
- Aesthetics of Dining Table Placement: Comprehensive Aesthetics Practice, focusing on the artistic principle of composition
- Food Garden: Sustainable Design of Permaculture
- Island Feast: Practice from Food Farming to Food Aesthetics
- Project-Based Inquiry: Research Methods

The learning modules, units, and learning activities were interconnected to enhance the continuous and interdisciplinary learning aspect of STEAM education. They provide flexible learning opportunities and can be adjusted appropriately in terms of the learning schedule, learning strategy, and learning space. To manage and to differentiate, I used three approaches, namely, spot, strap, and theme approaches, to facilitate teaching and learning. Illustrated in Figure 6.3, teachers and learners can arrange the spotted (e.g., single museum visit or half-day workshop), strap-shaped (e.g., multiple elective courses for whole semester), or thematic learning (e.g., project-based courses for several weeks, creating an e-portfolio to document students' learning progress) according to intended learning objectives and time.

Self-motivated, student-centered learning is the main teaching strategy for the natural aesthetic curriculum. Learners can study under the guidance of teachers or researchers, collaborate with parents and peers, or research and use digital resources to advance their learning that is centered on plants in Taiwan. The learning space included museums, schools, and communities. To reflect learners' unique life experiences, skill sets, and prior knowledge of their own regional culture, I divided the modules into elementary, intermediate, and advanced levels accordingly with age-appropriate content, instruction, and recommended duration for each learning activity.

3 Three STEAM Modules

In this section, I explain the content of the three STEAM modules and use students' work as examples to illustrate their learning of natural science, art, and selected aspects of Taiwanese culture. Emphasizing interdisciplinary and collaborative learning, students were able to acquire STEAM knowledge and enjoy practicing STEAM skills in their projects.

3.1 *Natural Science Inquiry*

The Natural Science Inquiry module took place at the NMNS. Students used the in-house i-NM app to facilitate self-learning online. In the learning activity called Diverse Plants in Taiwan: Learning Ecology at the Botanical Garden, students visited the largest greenhouse in Taiwan at the NMNS, containing plant collections from low-altitude ecosystems to tropical rainforests. By observing native plants like arbor trees, shrubs, herbs, and ferns, students understood the demographic of plants in Taiwan systematically and the interlocking relationship between the climate and natural environment.

In the learning activity called The Healing Power of Nature: Inquiry Learning at the Herb Garden, I introduced the connection between plants and the herb-based health supplements in Asian culture. The class opened in the herb garden with initial instruction on Chinese herbology (also known as "botanical knowledge"), including the collection, cultivation, and processing of medicinal plants; medicinal properties of the plants; prescriptions; and over-the-counter drug or supplement products. This activity enabled students to develop a foundational scientific understanding and appreciation of Asian medicine.

Researchers of natural science must have the ability to collect raw data such as images of nature and describe their meanings (Wood, 1979). Training for work in natural science necessarily includes the presentation of specific research evidence or inference results through images; therefore, the learning activity called Plant Artist: Scientific Illustration (Proportion) combined aesthetics and scientific knowledge to encourage students to gain skills used by natural science researchers. Specifically, students learned to produce observational hand-drawn scientific images. They chose and looked at plants carefully, based on the goal of their own research projects. They then considered the artistic principle of proportion to depict plants in detail.

Specimen production requires professional knowledge, tools, and techniques. Through the learning activity called Textures in Plants Collection: Specimen Production students gained the ability to identify a variety of flowering and fruiting plants and to distinguish morphological differences between woody and herbaceous plants. They recorded their observations using a plant identification table and learned to use the tools for plant collection, proceeding to produce their own plant specimens. In addition, they learned preservation techniques like shaping plant branches, pressing, air-drying, and sewing (see Figure 6.4).

The Natural Science Inquiry module allowed students to understand the relationships among Taiwan native plants, climate, and geographical environment; acquire knowledge of plant classification; and examine plants in the living environment. At the technical level they learned to collect plants and produce specimens with the assistance of experts; they recorded the plants with the accuracy of scientific drawing. Through guidance or the spontaneous use of museum learning resources, students learned the roles the plants play in human culture, connected the artistic principles of proportion and texture to natural science, and were motivated to complete the project-based inquiry of plants.

3.2 *Interdisciplinary Integration of Art and Natural Sciences*

In the module called Interdisciplinary Integration of Art and Natural Sciences, art teachers and natural science teachers collaborated to create an

FIGURE 6.4 A learner sew plant specimens

interdisciplinary learning environment. In the learning activity called Preserve Local Memory Color: Plants as Natural Dyes (Color) biology teachers connected students with information about plants provided by the NMNS. The content of learning focused on understanding the ecology of plants in schools and nearby communities. The goal was for students to gain knowledge of plant classification. They first observed coloration in plants to understand how sunlight affects the pigments in the plant cells. Then, they followed the steps to collect plants, prepare dyebath by conducting experiments to extract dyes from plants, apply tie-dye techniques to create geometric patterns, and dye fabrics in a pot with boiled water.

In the learning activity called Environmental Tactile Database: Handmade Paper (Texture) students collected plants fibers at their schools. They used microscopes to study natural fibers: the density, the thickness of the wall (supported by Sclerenchyma cells) to add strength, and the structure of plasma membrane. They discovered that depending on the length of plant fibers, they could produce diverse types of paper. Students learned to soak and cook plants to obtain proper amounts of fiber to make paper. Afterward, they learned about paper-making methods involving filtering paper pulps by scooping from the bottom to the top of the bin and dripping excess water back to the bin (See

FIGURE 6.5 Learner made a piece of paper in the shape of Taiwan by using plant fibers and stencils

Figure 6.5). Students further integrated aesthetic considerations and designs for their paper creations.

Students studied and applied the principles of aesthetics in the learning activity called Table Aesthetics: Comprehensive Aesthetics Practice (Composition). Charles Spence (2023), professor of experimental psychology at Oxford University, asserted that the way food is presented can affect people's taste expectations. Food plating, tableware in varied textures, the design of printed cloth mats, and the aesthetics of the table can all affect people's taste and dining experience. In the learning activities called Table Aesthetics the students integrated the arts and psychology by arranging plant-dyed cloth and handmade paper in an artistically pleasing way on the table to enhance taste expectations and eating experiences. Doing so, the students learned about the artistic principle of composition. Aiming at creating a feeling of harmony, they experimented with various compositions by using the geometric shapes and spaces generated by their plant-dyed cloth and handmade paper (Chang, 2007).

This module reinforces the knowledge gained in the previous activities about local plants. By adding the experimental process and aspects of biology, chemistry, and psychology, the students further applied principles of art and design, including color, texture, and composition, to foster their own aesthetics.

3.3 *Friendly Environment and Sustainable Ecology*

As a concluding activity the module called Friendly Environment and Sustainable Ecology focused on permaculture and ways to sustain our living environment. According to Bane (2012) and Holmgren (2003), permaculture is a portmanteau that combines "permanent," "agriculture," and "culture." It is an agricultural planning system that aligns with nature. Combining ecology, ethnology, and agronomy, the module introduced the following core values: looking after the earth, caring for people, and sharing. The activity called Food Garden: Sustainable Design of Permaculture integrated culinary arts and agricultural education. Students built eco-friendly vegetable gardens at school or at home, considering ways to preserve water and effectively managing their own gardens while avoiding harm to the environment. The module instructors also encouraged them to envision a career in permaculture and consider the roles and responsibilities they can assume to achieve a sustainable garden design.

Teaching the topic of plants with an interdisciplinary approach can motivate learners to explore the arts, the humanities, and ecology. In Island Feast: Practice From Food Farming to Food Aesthetics, guided by the module instructors, students brainstormed ideas for an island meal. They explored food and farming practices in Taiwan, looked at nutrition factors for a balanced diet, tended their gardens, harvested vegetables from their gardens, and used their harvest to create an island meal. Connecting to the knowledge obtained from Table Aesthetics in the previous module, they considered visual aesthetics in plating (see Figure 6.6).

4 Final Reflection

I believe that discovering and solving problems are essential competencies for STEAM learning. When designing this set of modules, I considered arts-based, STEAM, and inquiry-based pedagogical strategies (Kao, 2020). I also applied interdisciplinary and project-based learning principles (Stern et al., 2017). This approach helped cultivate students' research and problem-solving skills while exploring topics about plants, culminating in the opportunity to present their independent research on plants. Some studied mordants and their impact on coloration; others investigated permaculture and shared ideas for a sustainable garden. Finally, I believe that the success of this systematic learning process is a result of the collaboration of teachers, museum experts, and scholars, who applied information on plants found in cloud data, library materials, and local resources to develop relevant lesson plans and activities to motivate and enable learners to express and share their learning experiences.

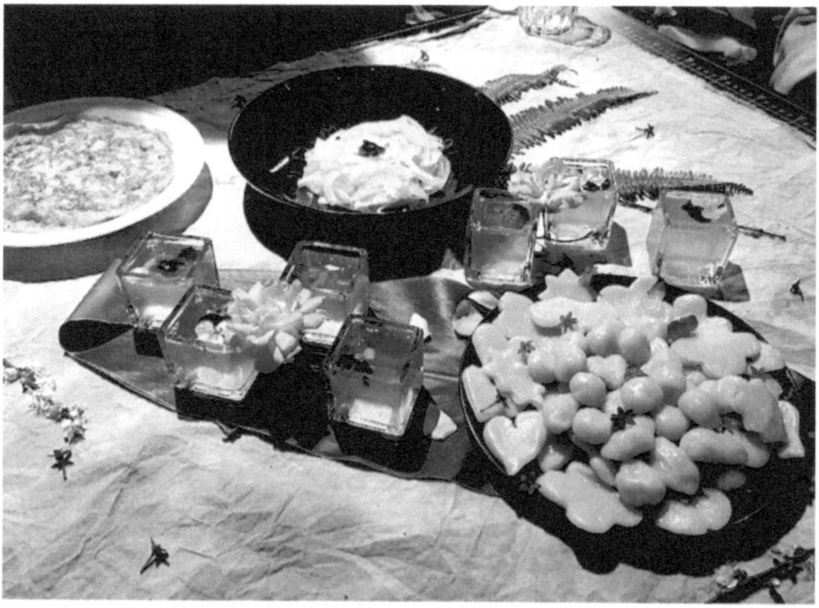

FIGURE 6.6 Student's example of an island feast

References

Bane, P. (2012). *The permaculture handbook*. New Society Publishers.

Chang, C. Y. (2007). Discussion on the creation using hand-made paper and media application fiber and replicated conception. *Visual Art Forum*, *2*, 59–72.

Holmgren, D. (2003). *Permaculture: Principles and pathways beyond sustainability*. Permanent Publications.

Hung, Y. S. (2012). The curriculum implication and development of aesthetic education in the cultural creativity era (文化創意世代中美感教育的課程圖像). *Journal of Educational Research and Development*《教育研究與發展期刊》, *8*, 61–85.

Kao, C. H. (2020). *A study of aesthetic experimental curriculum development and assessment in senior high school* [Unpublished doctoral dissertation]. National Taiwan Normal University.

Spence, C. (2023). Explaining visual shape-taste crossmodal correspondences. *Multisensory Research*, *36*(4), 313–345. https://doi.org/10.1163/22134808-bja10096

Stern, J., Ferraro, K., & Mohnkern, J. (2017). *Tools for teaching conceptual understanding, secondary: Designing lessons and assessments for deep learning*. Corwin.

Tsai, H.-C., & Liu, R.-L. (2020). Action study of community-based aesthetic education course design and practice for senior citizens. *Systemic Practice and Action Research*, *33*(2), 137–147. https://doi.org/10.1007/s11213-019-09484-x

Wood, P. (1979). *Scientific illustration: A guide to biological, zoological, and medical rendering techniques, design, printing, and display*. Wiley.

CHAPTER 7

Reflections on the Design of a STEAM Curriculum for 12- to 14-Year-Old Girls

Eliza Pitri, Maria Evagorou and Agni Stylianou-Georgiou

Abstract

In this chapter we reflect on the process of designing and implementing a STEAM curriculum for 12- to 14-year-old girls. The first cycle of the curriculum was designed and implemented in 2021 during a four-day summer school in Cyprus, involving a range of activities and problem-solving inquiry for designing uniforms for scientists in a variety of STEM professions. Based on findings, a new version was designed and implemented in summer 2022 to help students work on the problem of using color to support scientists in improving their well-being in their workspace. Lessons learned from the process of designing and implementing the curriculum included using a driving problem to guide activities, introducing activities that captured students' interests, explaining to students the design process and engaging them with it, and adding entrepreneurship activities. We highlight four pedagogical principles applied in the STEAM curriculum: learning with a purpose, gaining empowerment and agency, engaging in immersive learning, blending the STEAM disciplines through design thinking, and embracing uncertainty and risk taking. Keywords

Keywords

STEAM curriculum – engaging girls – design thinking – immersive learning

1 Introduction

The interconnection between the arts and sciences is historically rooted in the Renaissance. When Leonardo da Vinci crossed boundaries between disciplines as an active engineer, scientist, painter, sculptor, and architect, his notebooks replete with drawings from his studies of anatomy and astronomy as well as sketches of flying machines and shoes for walking on water (Kleiner, 2020). Da Vinci achieved balance between thinking rooted in the real world

 | DOI:10.1163/9789004714748_008

and dreaming about possibilities (Baggett & Macapia, 2020). His innate ability to balance empirical thinking leading to significant discoveries and inventive imagination resulting in the reconstitution of the known to create something new has inspired arts integration in education. Adding the arts to STEM education, however, requires a contemporary focus on more culturally relevant tasks than da Vinci's fashioning a silver lyre in the form of a horse's head or painting a fire-spitting monster inspired by the myth of Medusa. Sousa and Pilecki's (2018) work in both the public and private sector of education in the United States provided evidence that despite STEM teachers' reluctance to pursue STEAM due to concerns regarding time and resources, they and their administrators recognized that integrating the arts into STEM courses can increase student engagement in the learning process and awaken students' and teachers' creativity. Mae Jemison, an African American who developed a very early interest in anthropology, archaeology, evolution, and astronomy, became a physician, engineer, astronaut, and the author of several books as well as an art collector and a dancer. She called on educators during a TEDtalk to teach both the arts and sciences, both intuition and logic, as one way to create bold thinkers (Jemison, 2002). From da Vinci to contemporary STEM–STEAM role models like Jemison, the new renaissance of interrelating STEAM disciplines in education has led to pedagogical practices that allow students to learn subjects concurrently and demonstrate how in real contexts nothing is isolated (Hunter-Doniger, 2021).

Quigley and Herro (2019) recognized that adding the Arts to STEM aligns with all educators' commitment to adopt more humanitarian and equitable approaches to learning by considering the social and creative aspects of real-world problem solving. They highlighted the characteristics of STEAM teaching: problem solving, acknowledging multiple disciplines, and developing collaborative skills. As in the integration of any other field into education, STEAM lessons require meaningful engagement between disciplines, "creating a flow of ideas that ask for cooperative practices and vital input from different content areas" (McGarry, 2018, p. 31). Sousa and Pilecki (2018) analyzed the importance of the arts in the success of many scientists, mathematicians, and engineers, describing the skills that they use as scientific tools borrowed from the arts, which include the following:

- drawing on curiosity,
- observing accurately,
- perceiving an object in a different form,
- constructing meaning,
- expressing observations accurately,
- working effectively with others,

– thinking spatially, and
– perceiving kinesthetically.

Sousa and Pilecki referred to several studies, suggesting that integrating the arts into other content areas, such as STEM, has been shown to improve long-term memory, promote creativity, advance social growth, introduce novelty, and reduce stress. They concluded that these skills may not always be taught as part of STEM courses, but they are consistently taught in writing, drama, dance, painting, and music classes. Quigley and Herro add that creative skills can be employed in STEAM projects but are actually more evident today in the use of visualization and collaboration tools for video production and digital drawing that students readily practice outside school. STEAM projects differ from arts integration, which entails instruction through the arts; moreover, Watson (2016) pointed out that teaching STEAM lessons does not mean instruction of every discipline in every lesson, but it requires teachers to "regularly and authentically make connections between the disciplines whenever it makes sense to cross borders" (p. 8).

In STEAM education, the A of the arts adds an artistic, aesthetic, thoughtful, and creative lens in viewing engineering or design problems (Gettings, 2016). The European Union commissioned reports (e.g., European Commission, 2015), which are the driving force behind changes in national educational systems in Europe, also highlight the importance of moving from STEM to STEAM. In summary, STEAM is considered as an integrative approach to learning seeking to draw together various disciplinary perspectives (Yakman, 2010). The formal education programs currently offered in Cyprus favor learning that inducts students into a range of disciplinary forms but not necessarily into ways of seeing the connections among disciplines. In this chapter we focus on a reflective presentation of a STEAM curriculum delivered outside the formal school setting (i.e., in an "out-of-school" setting) for 12- to 14-year-old girls and describe the pedagogical principles underlying the activities as well as the participants' views on their implementation.

2 Why an Out-of-School STEAM Curriculum for Adolescent Girls?

Previous efforts focusing on increasing interest in STEM among girls included an emphasis on hands-on activities and the addition of creativity in STEM education (Calabrese-Barton et al., 2008). Girls' motivation to participate in STEM, in fact, increased after engaging in integrated STEAM activities. Merging the different STEAM disciplines has been considered as a way to engage female students with the content while at the same time supporting the development of critical thinking (Wieselmann et al., 2020). The curriculum presented in this chapter is part of a European-funded project called Empower Girls to Embrace their Digital and Entrepreneurial Potential, or GEM for short, that acknowledges the limited representation of women in STEM (Bilgin et al., 2022) and aims to address this issue by engaging girls in STEM from a younger age in order to increase their interest in the fields and empower them.

Our curriculum was designed for an out-of-school context encompassing extracurricular activities, summer school, and community activity. The choice of an out-of-school context for the STEAM curriculum was linked to findings from previous studies suggesting that such a context could enhance young students' interest in STEM (Heeg et al., 2022; Wieselmann et al., 2020) since a formal setting (i.e., a school) might limit the authenticity of the experience (Braund & Reiss, 2006). The flexible nature and lack of formal structure in the out-of-school context can allow integration of the STEAM disciplines. Teachers from various disciplines can teach together without a formal curriculum to follow. Furthermore, the nonstructured nature of out-of-school contexts can allow for activities that are longer in duration, support interaction with people from several disciplines and STEAM professionals, and make visits and field trips easier to organize, therefore providing a greater variety of educational experiences to students.

Some aspects of out-of-classroom STEAM activities reported as making the experience more authentic and enjoyable included the hands-on nature of activities, collaboration, creativity, and opportunities to exercise agency and to fail, all without the stress of being evaluated in a formal setting (Moore et al., 2014). Other reports from studies using STEM and STEAM out-of-school curricula emphasized students' preference for the out-of-school activities compared to those in school. Participation in such activities made it possible to collaborate with other students, provided them with the opportunity to ask questions of interest, and enabled them to learn from their mistakes (Heeg et al., 2022). Additionally, the evaluation of these activities highlighted their success in increasing students' motivation and STEM career aspirations (Kitchen et al., 2018); they empowered students (Habig et al., 2020) and showed potential to improve their knowledge and interest in learning (Chen et al., 2014).

3 The Context of STEAM Curriculum

Ten European countries participated in the GEM project, each country designing a curriculum to reflect its expertise and connect to its local context. Given the creative component that the A brings to STEAM, our research group in

Nicosia, Cyprus, decided to design a curriculum for integrated STEAM education instead of focusing only on STEM education. Our STEAM curriculum was designed for 10- to 14-year-old female students in Cyprus, a small island in Europe. Our roles during the summer school were diverse: We were involved in the design and instruction of most of the activities as well as the support of the process of preparing and presenting final products. However, for some of the summer school sessions (e.g., lab visits), the leading role in instruction was given to female role models or scientists.

The educational system in Cyprus is centralized with all schools for students aged 4–18 monitored by the Ministry of Education and Culture, which decides on the curricula and the books that will be taught in each age group. In middle school, students aged 12–15 focus on general knowledge in various subjects similar to primary education, but science is divided into chemistry, biology, and physics; and technology is added as a new subject. Students aged 15–18 may choose to attend high school where they must select a specialization (e.g., classical and humanities studies, STEM, finance); those who wish to follow STEM careers must choose the STEM specialization. Another option is a technical school that provides practical skills for students for a variety of professions (e.g., electrical engineering, construction, fashion design).

The arts are included in the curriculum in the form of architectural drawing, theatre, or drawing but are not compulsory, especially for high school students. The integration of the arts into STEM, however, can potentially support creativity in problem solving (Gettings, 2016) and support students with a range of interests and learning styles as well. The decision of our group was, therefore, to design a STEAM curriculum; but the very structured and centralized curriculum in Cyprus disallows the implementation of a STEAM curriculum in formal settings. Because of the affordances of out-of-school contexts, we designed the STEAM curriculum for a summer camp implemented in 2021 and 2022.

3.1 *STEAM Curriculum, Version 1*

Our research group started working on the first version of the STEAM curriculum in spring 2021. Based on findings from previous studies, we concluded that an out-of-school curriculum should include hands-on activities, group work and collaboration, creativity, agency, and opportunities to fail (Heeg et al., 2022). Given the emphasis of the GEM project on empowering and motivating young girls, we chose to include female role models as facilitators of activities and to engage students in entrepreneurial activities when possible. The first version of the STEAM curriculum was designed for a four-day full day (8 A.M.–3 P.M.) summer camp focused on the topic of color. From a pedagogical aspect, our research group tried to connect the topic of color to the different

STEAM activities each day (e.g., color in giant bubbles, color in the environment, color and indicators in the lab), provide a range of activities to the students (e.g., field study, inquiry-based activities, makerspace activities, lab visit), and continue with a problem-solving task related to color. The daily activities and tasks were designed in such a way to provide students with the content knowledge, skills, and experiences they would need to work on the main inquiry. The main inquiry was to design uniforms for scientists in a variety of STEM professions. The students worked on this main problem on the final two days, each group choosing a STEM profession and discussing in their groups how they would design the uniform and the reasons behind their choice. An overview of Version 1 of the curriculum as implemented in June 2021 appears in Table 7.1.

Version 1 of the curriculum was implemented in Summer 2021 with 20 students aged 12 to 14 and the participation of 12 female scientists and educators who worked together to implement the activities. During the implementation

TABLE 7.1 STEAM curriculum, Version 1

Day 1: Introduction and the Science of Color Refractions, reflections on prism or drop And a spectrum of colors would burst with a pop! Description: – Students were introduced to the Draw a Scientist activity and discussed stereotypes represented in their drawings with the rest of class. – Students engaged in inquiry-based experimentation, group work, and discussion of light and color (e.g., shadows, mixing colors). – Students watched a movie introducing them to the role of women in STEM. – Students formed groups and were asked to make giant bubbles large enough to accommodate a person. They also observed the bubbles in terms of color, consistency, and size. – Students discussed how bubbles are formed and what makes the colors on the bubbles iridescent.
Day 2: Color in the environment: Visiting the national park and environmental center Zoom in and look closely at any insects or birds you can find. As long as you are careful, we don't think they'll mind. Observe how their color patterns are like an intricate puzzle. So are colors in nature just all about dazzle? Description: – Students were introduced to the problem of the day: What is the role of color in the environment, and where can you spot iridescent colors in the National Park?

(*cont.*)

TABLE 7.1 STEAM curriculum, Version 1 (*cont.*)

- Students were given field study rules and the technological tools to use during the field study.
- Students then completed the following tasks: engage in field study in the National Park, present and discuss findings with female scientists, discuss the mechanisms of birds and bugs relating to the science of color with female scientists, and use materials from the environment to produce colors and discuss the technology of materials.

Day 3: Visiting a lab and meeting scientists

Some think that the spectrum holds only magic and mystery,
But the science of color developed throughout history,
Provides information for doctors and chemists in labs,
And offers all scientists indisputable facts.

Description:

- Students investigated the question: How can we use color in a laboratory? They worked in groups in the science lab, mentored by a female scientist to understand the importance of color in chemistry and utilized color to find out blood type.
- Students interviewed a female scientist: They met online a member of the Cypriot epidemiological team and had the opportunity to interview her about her career, everyday life, and difficulties that she encountered as a female scientist and science communicator.
- Students in groups chose a scientist and discussed a new uniform design.
- Students made drawings and calculations using the design process to prepare outlines of uniforms.

Day 4: The makerspace experience and working on the final project

So just choose a scientist and get in their positions!
What would you do and what would you wear under those conditions?
Think about colors and the role they would have
In a uniform design elevated to stardom and above.

Description:

- Students visited a local STEM makerspace center and learned about entrepreneurial skills in STEM and basic rules as well as the technologies and tools available for them to use (e.g., 3D printer, programming).
- Students prepared their final project: They applied what they learned about color and STEM to create a scientist's uniform for their wooden mannequin, considering the use of color and whether it had a particular function in their design (see Figure 7.1).
- Students planned the presentation of their designs as a way to promote their final product and concluded with a presentation.

FIGURE 7.1 Photo from Day 4: final product created by the groups (uniforms for four people working in STEM)

we collected feedback from the students, summarized in the following points: (1) Students were excited about designing uniforms for the scientists, but they did not have the necessary practical skills, namely sewing and cutting fabric. (2) Students were excited about visiting a science lab and working with a scientist but would like more time to spend in the lab. (3) Students asked for more time with technology and technological tools and talked about the need to learn programming and engage with 3D printing. (4) Students asked for less theoretical sessions, more practice, and more time for play.

The educators who implemented the activities with the students observed the following: (1) The activities were not truly integrated. The connection among STEAM subjects was "forced" by the teaching team and was prelinked to the topic of color. The activities did not bring a variety of disciplines together to address a problem. (2) Students needed support in the design process. Almost all students went from discussing the idea to trying to make a working prototype without first discussing alternative solutions, exploring pros and cons of their solution, and creating nonworking prototypes. (3) Students lacked practical skills (e.g., measuring and sewing) needed to design a uniform.

3.2 *STEAM Curriculum, Version 2: Learning from Practice*

Observations, comments, and reflections from the first implementation of the STEAM curriculum were useful for our research group to support the preparation of a new, 5-day, version of the STEAM curriculum implemented in June 2022. The main objectives of the second curriculum, presented in Table 7.2, were to

- introduce young students to STEM disciplines,
- help students understand STEM stereotypes assessed through the Draw a Scientist task,
- bring students in contact with female STEM professionals as role models,
- help students appreciate various aspects of STEM professions,
- engage students with skills linked to STEM and STEAM (e.g., creativity, programming, inquiry-based learning, problem solving), and
- engage students in problem solving and entrepreneurial activities.

TABLE 7.2 STEAM curriculum, Version 2

Day 1: Introduction and the science of color Can science take me to the moon and back? Yes, but STEAM can do way more than that! Description: – Students learned the aims of the summer school program and engaged in ice-breaking activities to meet one another. – Students completed the Draw a Scientist activity and discussed stereotypes represented in their drawings with the rest of class. – Students were immersed in virtual reality and augmented reality environments in which they could manipulate colors at the virtual reality lab at the University of Nicosia. – Students engaged in discussion with female role models: They met an astrophysicist and discussed her professional life and journey to where she is today. – Students were introduced to the problem of the week: How to use color to improve scientists' well-being in their workspace.
Day 2: Color in the environment: Visiting the national park and environmental center Whether in office, classroom, or lab, An appropriate space could contribute to work done. Description: – Students were introduced to the problem of the day: What is the role of color in the environment, and where can you spot iridescent colors in the National Park? – Students learned the field study rules and the technological tools used during the field study. – Students completed the following tasks: engage in field study in the National Park, present and discuss findings with female scientists, and design and present a sustainable school using programmable building blocks (LEGO® Education WeDo 2.0).

(*cont.*)

TABLE 7.2 STEAM curriculum, Version 2 (*cont.*)

Day 3: Using color in the lab and the mathematics of color Cynthia, Gladys, Juliana, Mae, or Hayat, Learn what it takes to be a STEAM rising star. If you would like, you can be the next one. Follow female STEAMers just because you can! Description: – Students visited the human biology lab at the University of Nicosia and learned about basic rules and how to use color in lab work. – Students worked in the biology lab with a female immunologist and used colored indicators in order to identify their own blood type. – Students learned the role of women in STEM through an interactive presentation of women in STEM fields: Cynthia Breazeal, Gladys West, Mae Jemison, Juliana Rotich, and Hayat Sindi. – Students participated in a workshop on programming, LEGO® Education WeDo 2.0, using light and color as the problem. – Students watched the movie *Hidden Figures* (Melfi, 2016) featuring women and mathematics.
Day 4: The design thinking process and the problem Think about what female scientists do. And consider how context could make their work bloom. Come up with a prototype, discuss, and produce. Description: – Students engaged in guided group work on design thinking for solving the problem: How can we use color to improve scientists' well-being in their workspace? – Students worked in groups to design a product to address the problem.
Day 5: Entrepreneurial skills in STEM Finally, present your group's suggestion. And visually explain your decisions and direction. Female scientists deserve our attention, And your creations will all be worth mention! Description: – Students worked in groups to produce a prototype of the product related to the problem and a logo for it. – Students met female role models (interface designer and software designer), who presented their work and explained how to pitch a product. – Students presented and pitched their final products and logos, which were evaluated by female role models.

Each day students were presented with a set of activities designed to provide them with the necessary skills, knowledge, and experiences to help them work on the main problem. The main problem introduced on Day 1 was "How can we use color to improve scientists' well-being in their workspace?" The students,

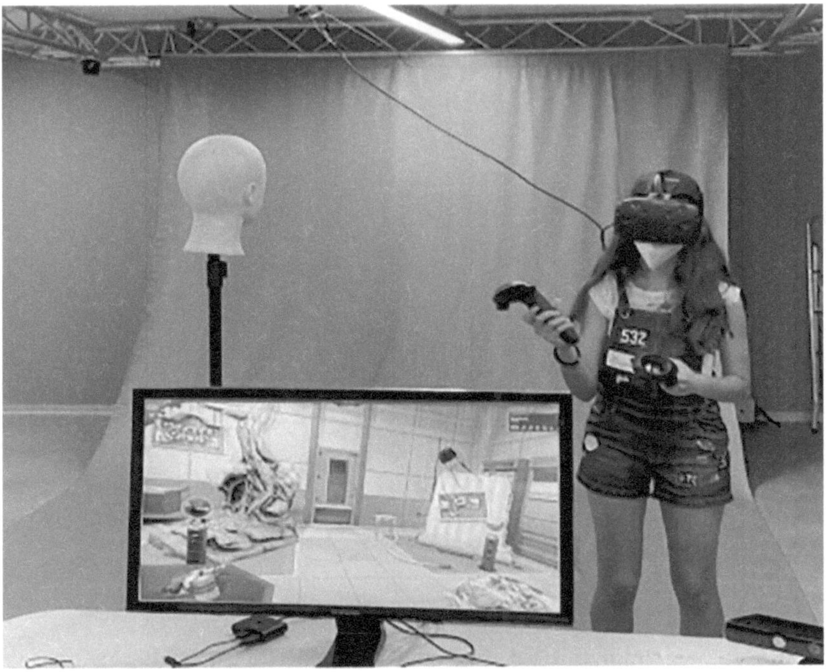

FIGURE 7.2 Photo from Day 1: exploring a lab in a virtual environment

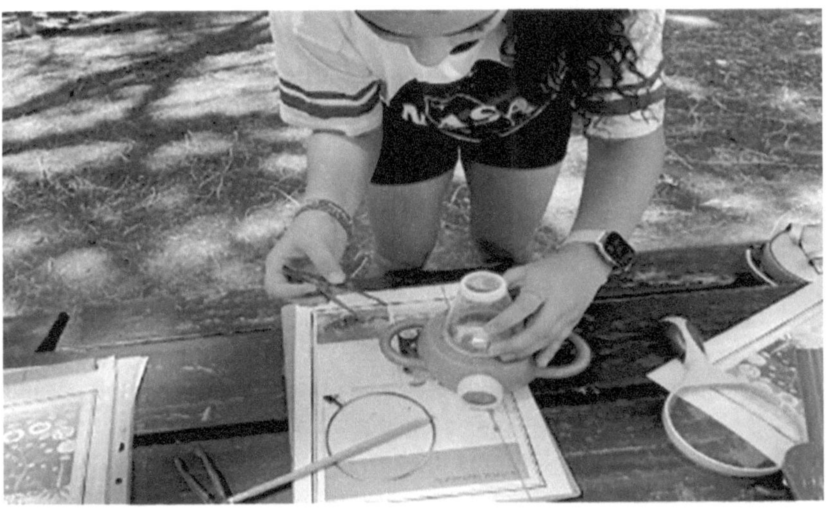

FIGURE 7.3 Photo from Day 2: observing organisms collected from the soil at the park

in their groups, worked on the main problem during the final two days of the implementation.

Toward the end, each one of the groups designed a product to address the problem. For example, one group designed a pillow with an LED light

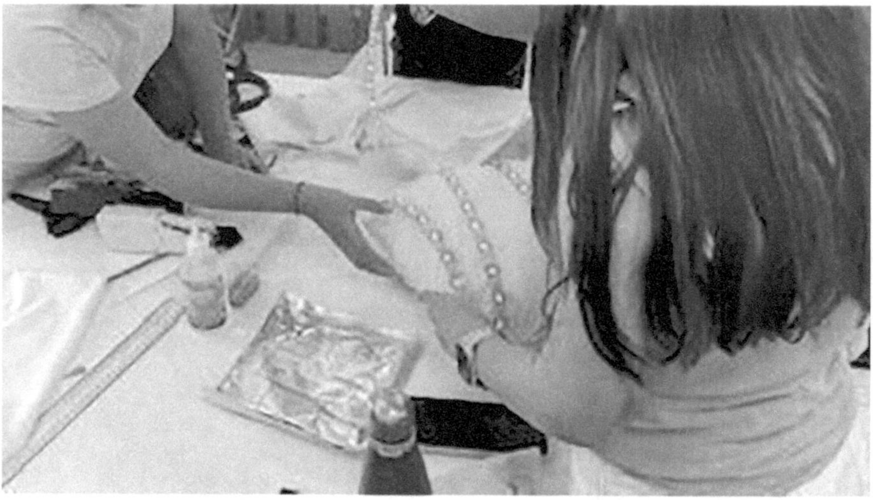

FIGURE 7.4 Photo from Day 5: working on the prototype of the product, a pillow that can change color

that could change color to help the scientist relax when working in the lab (see Figure 7.4). Another group designed a robot car, programmed to transfer materials to the scientists; in addition, it was also programmed to change color after one hour to remind the scientist to take a break.

The second STEAM curriculum was implemented with 27 girls, aged 12 to 14 years. Data about students' interest in STEM before and after the summer school were collected using pre- and postclass questionnaires. Findings indicated that students' interest in learning about STEM increased after engaging with the curriculum. The most important change in their views indicated that after engaging with the curriculum, most of them reported that they understood how STEM can help them in their future careers and their everyday lives. Furthermore, students were asked to provide feedback on the STEAM curriculum and the activities. The students commented very positively on the fact that they worked in groups most of the time, with student 9 reporting as follows: "My favorite activity was working on the product that we had to prototype, and I enjoyed it because we worked as a group and had fun at the same time."

The students also enjoyed a range of activities that included visiting labs (e.g., virtual reality lab, biology lab) and working with scientists, engaging in outdoor activities, and watching movies. Student 21 reported, "I really enjoyed the virtual reality games on the first day because it was my first experience with something like this. I also liked the day at the park because I like being outside." Student 13 stated, "I really enjoyed working in the biology lab, and I got excited because I learned about my blood type by doing the experiment myself."

Students were asked to work on their product, prototype it, create a logo for it, and pitch it during the final two days of the implementation of the curriculum, tasks that they all reported as positive aspects of the curriculum; but 13 of the 27 students reported frustration about failure, especially when they had to rework some steps of the process when their products did not work. Groups worked on a variety of products and consequently encountered a range of types of failure. For example, one of the groups designed the prototype for a pillow that could change color. Student 6 said, "[I was] really excited and happy about gaining new skills and knowledge, but I got frustrated when our prototype was not working, and we had to build it again." Feelings of frustration were also reported during the implementation when products did not work even though problems were later resolved.

4 The Pedagogical Principles of the STEAM Curriculum

Based on a combination of our research group's personal reflections and the literature, we highlight four pedagogical principles in this section, comparing the use of these principles in the two versions of our STEAM curriculum. We found that they especially improved students' learning in the second version of the STEAM curriculum.

4.1 *Learning with a Purpose, Empowerment, and Agency*

Designing the curriculum, we aimed at engaging the teenaged girls in a learning environment that "ignited" a creative spark as they learned about STEAM. In this environment students were active in the learning process and teachers were coaches, guiding the understanding of concepts and making connections to students' everyday experiences (Amabile, 1996). We chose color as the common main theme in all learning activities for both versions of the STEAM curriculum. The first curriculum showed us that setting a purpose for learning was important but insufficient. During our implementation of the second STEAM curriculum, we attempted to create a learning environment that encouraged students to be cognitively engaged in meaningful inquiry and take responsibility for their own learning. For example, students in the first STEAM curriculum completed small tasks to learn about the science concepts related to color, but when they created the scientist's uniform, they did not necessarily apply that knowledge to make choices that reflected the knowledge gained. In the second STEAM curriculum, the main problem— using color to improve scientists' well-being in their workspace—guided the design of all activities across the curriculum. Through the presentation of the problem, we aimed to empower students to engage in creative processes and take agency in the work. When learners gain a sense of their own agency and self-expression (brainstorming, generating questions, finding, and applying information), they engage in processes producing knowledge in meaningful ways (Blumenfeld et al., 2006).

4.2 *Immersive Learning*

In our first attempt to design the curriculum, we tried to incorporate a "wow factor" into the complex challenges that would engage the teenaged girls (e.g., entering a giant bubble, presenting scientists' uniforms in a fashion show). In this way students were motivated but not overwhelmed by the challenge assigned to them, and they were not frustrated by a demanding task even though they were eventually able to complete it. However, the wow factor proved to be a double-edged sword since students failed to understand how the activities in the curriculum were integrated even though they were immersed in the learning activities. Csikszentmihaly's (1990, 1993) research focusing on individuals' optimal experiences indicated that such experiences are associated with feelings of deep enjoyment and happiness, which he defined as "flow." He suggested that flow occurs when individuals become absorbed while facing challenges that are neither too difficult nor too easy. For flow to exist in a learning environment, teachers should shape learning activities at a challenging level but not above the levels of students' ability. In order to create this immersive experience, we revised our curriculum in order to guide students to make connections to the problem they explored, following a design thinking approach.

4.3 *Blending the STEAM Disciplines through Design Thinking*

Because integrating knowledge could not be achieved through a task-based approach (i.e., a variety of tasks to learn about color) in the first STEAM curriculum, we decided to add a design thinking approach (Razzouk & Shute, 2012) to the second STEAM curriculum so that activities would not be conducted or construed as disconnected tasks. The girls were engaged in a cyclical process of four nonlinear stages of design thinking: empathize, define, ideate, and prototype. Through this process they could gain an insight into the problem, identify user needs, experiment, create and prototype models, gather feedback, and redesign. Upon presentation of the model, the girls engaged in information gathering and developed a solution to a design problem. The design thinking framework proved to be a valuable tool because it guided them to develop strategic ideas for relevant implementation, applying their varied knowledge and skill levels; to vitally connect to what they were learning and doing; and to enact their learning by producing something rather than simply receiving knowledge.

During the first STEAM curriculum, when the girls engaged in a design challenge at a local makerspace center, they seemed to be overwhelmed by the range of tools and materials available when designing the scientists' uniforms. Through "tinkering" (Resnick & Rosenbaum, 2013), the girls followed an unplanned approach to experimenting with materials. Even though they were able to explore and try new things through a playful, experimental, and iterative style of engagement, they were not able to reassess their goals, including exploring new paths and imagining new possibilities. During the second STEAM curriculum, before they started creating a prototype of the idea, which was collectively generated, instructors encouraged them to follow the stages of design thinking to generate questions and think of the users; furthermore, instructors offered guidance on tools (e.g., robotics toolkits) only when learners needed them to develop their ideas. Combining tinkering and design thinking thus proved to contribute to STEAM learning as both approaches provided students with a framework for thinking about their own practice.

4.4 *Embracing Uncertainty and Risk Taking*

In both the first and second versions of the STEAM curriculum, the instructors and female scientists encouraged the girls to think and act like designers in order to solve an ill-structured problem (Reed, 2016). In ill-structured (design) problems the goal and the initial and intermediate stages of the problem are incompletely specified. Designing a scientist's uniform or improving scientists' well-being in their workspace involved uncertainty as there are many possible solutions to each problem. This, in turn, presented opportunities for the girls to exercise creative thinking, take risks, and embrace uncertainty. In the second version of the STEAM curriculum, the instructors allowed the girls to identify problems of interest to them, look for their own solutions, share ideas, and receive feedback. This is what Beghetto (2018, 2021) described as "creative openings" in the curriculum. Like other researchers (Finke et al., 1992) Beghetto (2019) has argued that creative expression does not exclude predetermined limits or criteria. Instead, creative openings provide opportunities for students to engage in innovative endeavors as they work in a problem space that possesses multiple solutions and uncertainty about which concepts, rules, and principles are necessary for the solution.

By opening up the curriculum to creative expression, we allowed the girls to deal with uncertainty (e.g., facing unexpected twists and turns) and to seek out unique ways to demonstrate their understanding of content knowledge and skills across the curriculum. During both implementations we noticed that embracing uncertainty engenders possibility and a "can-do" attitude that can empower students to take ownership of their own actions and self-expression (Craft, 2001). Based on this view, the problem space can create new possibilities for creative expression (Pitri & Stylianou-Georgiou, 2022), pushing learners to think and act with creative confidence and cultivate awareness of their personal abilities and domain knowledge, judging whether—based on the conditions of the environment in which they learn—it is worth taking a risk (creative metacognition) for creative action (Beghetto, 2017).

5 Conclusion

Designing and implementing a STEAM curriculum for an out-of-school context includes creative aspects and has been an inspiring process for our group. The unstructured nature of the out-of-school context in combination with the goal of integrating the STEAM disciplines led our teaching and research group to explore ways to engage the students. Our first concluding thought about the process of designing a STEAM curriculum relates to integrating the STEAM disciplines. This is an important part of a STEAM curriculum, and we recognize that integration should be presented through a problem that guides the curriculum and is shared with the students from the outset. Students can then work collaboratively and bring together skills and knowledge from all disciplines to provide a solution. What supported a more successful implementation of the curriculum in terms of integration was the introduction of the design thinking process as a way not only to scaffold students to think about the problem and the solution from multiple perspectives but also to prepare for their presentations.

Our second thought about the process of designing a STEAM curriculum relates to the involvement of others in the process, especially professionals who bring their skills and expertise to support the integration of the STEAM disciplines. Working with experts in their respective fields of study has provided students with authentic experiences and an understanding of the skills and processes necessary in those disciplines. This experience also provided them with the context—information about the workplace of STEM professionals—that was required to provide a solution to the problem.

Our final thought is that an out-of-school STEAM curriculum requires a balance of learning and fun. The open structure of the problem provided the opportunity for students to work on solutions relevant to the problem but also meaningful to them. Working on a solution was driven by their own interests, self-initiated, and relevant to them, thus fostering student agency.

6 Acknowledgments

This chapter is based on work done in the GEM project, cofunded by the European Union under Grant No. LC-01380173. The EU/EC is neither responsible for the content nor liable.

References

Amabile, T. M. (1996). *Creativity in context*. Westview Press.

Baggett, A., & Macapia, C. (2020). Vending delight with elementary students. *Art Education, 73*(6), 37–42. https://doi.org/10.1080/00043125.2020.1785798

Beghetto, R. A. (2017). Creativity in teaching. In J. C. Kaufman, V. P. Glăveanu, & J. Baer (Eds.), *The Cambridge handbook of creativity across domains* (pp. 549–564). Cambridge University Press.

Beghetto, R. A. (2018). *What if? Building students' problem solving skills through complex challenges*. Association for Supervision and Curriculum. (ED586941) ERIC.

Beghetto, R. A. (2019). *Beautiful risks: Having the courage to teach and learn creatively*. Rowman & Littlefield.

Beghetto, R. A. (2021). Creativity in K–12 schools. In R. J. Sternberg & J. C. Kaufman (Eds.), *Creativity: An introduction* (pp. 224–241). Cambridge University Press.

Bilgin, A. S., Molina Ascanio, M., Milanovic, I., Kirsch, M., Beernaert, Y., Scicluna Bugeja, D., Noriega, M., Farrugia, J., Evagorou, M., Molina, P., Kapoor, K., Malmberg, B., Trullàs, M., Pedralli, M., Neuberg, C., Koliakou, I., Magid-Podolsky, S., Herrero, B., Christou, E., … Gras-Velázquez, A. (2022). STEM female leaders—The way forward to reduce the gender gap in STEM fields. *SCIENTIX*. https://kavitakapoor.org/Bilgin_et_al_2022_STEM_Female_Leaders_STNS.pdf

Blumenfeld, P. C., Kempler, T. M., & Krajcik, J. S. (2006). Motivation and cognitive engagement in learning environments. In R. K. Sawyer (Ed.), *The Cambridge handbook of the learning sciences*. Cambridge University Press.

Braund, M., & Reiss, M. (2006). Towards a more authentic science curriculum: The contribution of out-of-school learning. *International Journal of Science Education, 28*(12), 1373–1388. http://dx.doi.org/10.1080/09500690500498419

Calabrese-Barton, A. C., Tan, E., & Rivet, A. (2008). Creating hybrid spaces for engaging school science among urban middle school girls. *American Educational Research Journal, 45*(1), 68–103. https://doi.org/10.3102/0002831207308641

Chen, C., Tomsovic, K., & Aydeniz, M. (2014). Filling the pipeline: Power system and energy curricula for middle and high school students through summer programs. *IEEE Transactions on Power Systems, 29*(4), 1874–1879. http://dx.doi.org/10.1109/TPWRS.2013.2293752

Craft, A. (2001). Possibility thinking: From what is to what might be. In A. Craft, B. Jeffrey, M. Leibling (Eds.), *Creativity in Education* (pp. 45–61). Continuum.

Csikszentmihalyi, M. (1990). *Flow: The Psychology of Optimal Experience.* Harper & Row.

Csikszentmihalyi, M. (1993). Activity and happiness: Towards a science of occupation. *Journal of Occupational Science, 1*(1), 38–42. https://doi.org/10.1080/14427591.1993.9686377

European Commission, Directorate-General for Research and Innovation. (2015). *Science education for responsible citizenship: Report to the European Commission of the Expert Group on Science Education.* Publications Office. https://data.europa.eu/doi/10.2777.12626

Finke, R. A., Ward, T. B., & Smith, S. M. (1992). *Creative cognition: Theory, research, and applications.* MIT Press.

Gettings, M. (2016). Putting it all together: STEAM, PBL, scientific method, and the studio habits of mind. *Art Education, 69*(4), 10–11. https://doi.org/10.1080/00043125.2016.1176472

Habig, B., Gupta, P., Levine, B. & Adams, J. (2020). An informal science education program's impact on STEM major and STEM career outcomes. *Research in Science Education, 50*, 1051–1074. https://doi.org/10.1007/s11165-018-9722-y

Heeg, D., Smith, T., & Avraamidou, L. (2022). Children's experiences and self-identification with science in the context of an out-of-school STEM program. *EURASIA Journal of Mathematics, Science and Technology Education, 18*(4), Article em2091. https://doi.org/10.29333/ejmste/11888

Hunter-Doniger, T. (2021). Forming artist/scientist habits. *Art Education, 74*(2), 16–21. https://doi.org/10.1080/00043125.2020.1852376

Jemison, M. (2002, February). *Teach arts and sciences together* [Video]. TED Conferences. https://www.ted.com/talks/mae_jemison_teach_arts_and_sciences_together?language=en

Kitchen, J. A., Sonnert, G., & Sadler, P. M. (2018). The impact of college- and university-run high school summer programs on students' end of high school STEM career aspirations. *Science Education, 102*(3), 529–547. https://doi.org/10.1002/sce.21332

Kleiner, F. S. (2020). *Gardner's art through the ages: A global history* (16th ed.). Cengage Learning.

McGarry, K. (2018). Making partnerships with STEAM. *Art Education, 71*(2), 28–34. http://dx.doi.org/10.1080/00043125.2018.1414535

Melfi, T. (Director). (2016). *Hidden figures* [Film]. 20th Century Fox.

Moore, T. J., Stohlmann, M. S., Wang, H.-H., Tank, K. M., Glancy, A. W., & Roehrig, G. H. (2014). Implementation and integration of engineering in K–12 STEM education. In S. Purzer, J. Strobel, & M. E. Cardella (Eds.), *Engineering in pre-college settings:*

Synthesizing research, policy, and practices (pp. 35–60). Purdue University Press. https://doi.org/10.2307/j.ctt6wq7bh.7

Pitri, E., & Stylianou-Georgiou, A. (2022). *Boundaries in artmaking—Windows to expression.* Paper presented at the Annual Meeting of the American Educational Research Association, San Diego, CA, USA.

Quigley, C. F., & Herro, D. (2019). *An educator's guide to STEAM: Engaging students using real-world problems.* Teachers College Press.

Razzouk, R., & Shute, V. (2012). What is design thinking and why is it important? *Review of Educational Research, 82*(3), 330–348. https://doi.org/10.3102/0034654312457429

Reed, S. K. (2016). The structure of ill-structured (and well-structured) problems revisited. *Educational Psychology Review, 28*, 691–716. https://psycnet.apa.org/doi/10.1007/s10648-015-9343-1

Resnick, M., & Rosenbaum, E. (2013). Designing for tinkerability. In M. Honey & D. Kantor (Eds.), *Design, make, play: Growing the next generation of STEM innovators* (pp. 163–181). Routledge.

Sousa, D. A., & Pilecki, T. (2018). *From STEM to STEAM: Brain-compatible strategies and lessons that integrate the arts.* Corwin.

Watson, A. D. (2016). Revving up the STEAM engine. *Art Education, 69*(4), 8–9. https://doi.org/10.1080/00043125.2016.1178032

Wieselmann, J. R., Roehrig, G. H., & Kim, J. N. (2020). Who succeeds in STEM? Elementary girls' attitudes and beliefs about self and STEM. *School Science and Mathematics, 120*(5), 297–308. https://doi.org/10.1111/ssm.12407

Yakman, G. (2010). What is the point of STEAM? A brief overview. *Academia.* https://www.researchgate.net/publication/327449281_What_is_the_point_of_STEAM-A_Brief_Overview

CHAPTER 8

Application of Data Visualization in Hakka Cultural Education in Taiwan

Shyh-Shiun Shyu and Yui-Chih Wu

Abstract

In this chapter we share our use of a STEAM-infused curriculum to enable children to understand the cultural, historical, and family data of the Hakka ethnic group at a 2021 online summer camp. Data visualization can help children connect with their families closely, understand ethnic groups deeply, compile information objectively, and interpret stories based on and through raw data. We concluded that data visualization not only helps children explore family emotions and the collective memory of their ethnic group but also gives them an essential dose of data literacy.

Keywords

STEAM – data visualization – Hakka culture – ethnic groups

1 Introduction

The key task of the Taipei Hakka Culture Foundation (hereinafter, the "Foundation") in Taiwan is to maintain the legacy of Hakka culture and education. The Foundation frequently uses children's summer camps to achieve this goal. In 2021 the Foundation introduced a contemporary trend in art education—data visualization—to facilitate children's understanding of Hakka culture and to guide them to explore the historical migration of Hakka families and clans. Through the visualization process, cultural, historical, and family data can be converted into a variety of forms, such as charts, graphs, and maps, artworks that can be used by the children to project their understanding of Hakka culture. Ensuring that the information behind the data is interpreted and understood more intuitively, the application of data visualization approaches for teaching Hakka culture not only offers new perspectives for the interpretation

 | DOI:10.1163/9789004714748_009

of data but also reaffirms relationships among people, events, things, and the environment while creating impactful knowledge (Urist, 2015). Data visualization, that is, a creative presentation of information, is not the end goal of a particular teaching process but aims to expand learners' thinking about what constitutes information design, the process of data analysis, and what the data output looks like (D'Ignazio & Bhargava, 2018). Based on these considerations, we integrated data visualization into a summer camp curriculum.

This chapter highlights learning about Hakka culture in an online summer camp in 2021. The Hakka descendants who participated in the camp experience procured data pertaining to their family history through discussion with their parents; they then represented their family history using various colors. While doing so, they infused the data visualization process with Hakka traditional imagery, such as hand-wrapped silk flowers, symbolizing their ancestors' cultural heritage and hopes for the future. They used color to convey the landscape topologies. The students gained a new understanding of the experiences of their Hakka ancestors when they initially arrived and eventually dispersed throughout Taiwan to work in agriculture. These methods increase the likelihood of students' familiarity with the manner in which Hakka populations coalesced in the past, the history of the movement of their own families and clans, and the distribution of Hakka industries. Information regarding the Hakka culture can be transmitted within the Hakka population through stories that can also be understood by non-Hakka populations.

2 A Brief History of Hakkas in Taiwan

Over the course of the past three centuries, Taiwan was ruled by the Spanish and Dutch empires, the Qing Dynasty, and Japan among others. Taiwan's current population comprises four large ethnic heritage groups: Southern Min (Hoklo), who migrated from the southern Fujian province in China; Hakka from various provinces in southern China; later 20th-century migrants from mainland China as a whole (the so-called *waishengren*); and indigenous people, whose ancestors have been in Taiwan all along. These four groups have created the primary threads in Taiwan's rich cultural tapestry; however, during the early years of the Qing Dynasty (1644–1911), Taiwan's population was split along ethnic lines, and group conflicts were frequent. The Hakka people did not disclose their identities or reveal their language and customs, so distinguishing them from other ethnic groups was difficult. The cultivation and transmission of Hakka culture were, therefore, challenging (Chiou, 2012).

The ethnic composition of Taipei's population differs from that of Taiwan as a whole, Hoklo people accounting for the majority, followed by *waishengren*; Hakka and indigenous people are considered minority groups (Yang, 2004). Like other groups in Taipei, Hakka people trace their origins to other cities in Taiwan. The island covers a small land area, but its unique geography and natural environments have created heterogeneous living conditions in various locales. The Hakka ancestors in other parts of Taiwan could adapt to the challenges they faced in unique ways, which were transmitted to later generations, thereby becoming the unique characteristics of Hakka families. Thus, we can label the entire group of people as "Hakkas," but the Hakka population in Taipei shows nuances because they are derived from Hakka settlements in many areas of Taiwan. The Hakka subcultures in Taipei stand as a microcosm of the subtle intermingling of the ethnic groups and cultures within the borders of Taiwan.

Although Taipei has numerous political and economic advantages as well as abundant cultural resources, promotion of Hakka culture in Taipei still remains a challenge because of its history of intentionally subduing cultural distinctions. Furthermore, the Hakka people had to learn the languages (Mandarin and Hoklo) of the dominant ethnic groups in order to earn a living and adapt to the urbanized social environments while concealing their own language and culture (Huang, 2012). This drive toward suppression and "passing" has recently emerged as an important research topic in Hakka cultural studies, clarifying the challenges in Hakka cultural education in Taipei and making the following a pertinent question: How can non-Hakka students be introduced to the Hakka culture while avoiding cultural appropriation?

Because the Hakka people migrated from their original homelands in China, a straightforward approach to stimulating a sense of Hakka identity is to reproduce symbols used in traditional Hakka culture. Cultural artifacts, texts, and symbolic imagery associated with Hakka people include the following:

- Cultural products: camphor, blue dye, *lui ca* (pounded tea), and oil-paper umbrellas;
- Hakka foods and culinary elements: tea; pork offal with shredded ginger; Hakka flat noodles; Hakka stir-fry; Chinese mesona (*Platostoma palustre* or 仙草*xienco*), a type of mint; salty, intensely aromatic, strongly flavored dishes; fermented foods; and distinct pickling methods;
- Festivals: 天穿 *Tien-chon*, Tung Blossom Festival, and 義民 Ngi-min Festival;
- Designs: Tung blossoms, Hakka floral fabrics; and
- Farming: Seasonal planting.

3 Curricular Framework 1: Data Visualization and STEAM Approaches

Data visualization helps students to integrate skills, connect knowledge, and build visual literacy (Cooper, 2014). The process of making data visualizable involves turning *data* into *information*, which can be used to impart *knowledge*. The process of turning knowledge into wisdom helps students integrate disparate concepts into wisdom that carry enlightening meaning (Letouzé et al., 2015); however, several challenges emerge when transmitting and promoting Hakka culture in Taiwan. First, the history of the Hakka population is vast, deep, complex, and seemingly very distant from children's everyday lives and experiences. Second, modern children study core school subjects, leaving almost no time for extracurricular topics like ethnic culture; consequently, Hakka culture studies are often reduced to rote duplication of cultural symbols (e.g., Tung blossoms, Hakka floral fabrics, oil-paper umbrellas).

Considering the value of data visualization, Foundation leaders decided to introduce data visualization at a children's summer camp, effectively using it to present Hakka culture and their family and clan histories through calculation, cultural symbology, and artistic creation, facilitating understanding of the information acquired. Cultivating students' general literacy in this manner made it easier to connect the dots between concepts in areas like art, society, culture, and history. The composite meaning facilitated cultural learning experiences (Cooper, 2014; National Coalition for Core Arts Standards, 2014).

We also considered the application of data visualization in our summer camp as a form of STEAM education, which includes science, technology, engineering, the arts, and mathematics. The "A" includes topics derived from art, design, and the humanities. While planning our summer camp curriculum, we focused on Hakka culture as our core concept, around which we developed topics and ideas like history, culture, knowledge about naturally produced and manmade items, intergenerational comparative conversations, arithmetic, and artistic creation. These topics covered the arts and mathematics aspects of STEAM. Art is perceived as a way of interpreting STEM content, which includes data visualization, that is, the ability to transform data and information through visualization skills (Cooper, 2017; Liao, 2019). We, therefore, guided students to explore the histories of their families and clans, using arithmetic to understand the Hakka culture and then make and present art based on their findings. Our aim behind the design of the STEAM-oriented summer camp curriculum was to enable students learn to observe, select, and analyze content related to the themes from the raw data, then use their creativity to think about, create, and communicate the findings effectively (Bertling et al., 2021; Cooper, 2014). These were the educational goals we wanted to achieve.

4 Curricular Framework 2: Ecological System Theory

The Foundation, which works to promote children's ethnic and cultural education, was inspired by the ecological systems theory created by the developmental psychologist Urie Bronfenbrenner (1917–2005), who stated that children are influenced by at least four ecological systems. The *microsystem* includes those interactions in which the individual directly participates, such as the nuclear family, babysitters, grandparents, teachers, and peers. The *mesosystem* consists of interactions created between two microsystems, such as connections and contacts between a family and a school or between peers and the community. The *exosystem* consists of external environmental factors that indirectly influence the individual, such as the educational tendencies of families and schools, the use of resources by communities, and social organizations and services. The *macrosystem* includes those factors that indirectly influence the individual, such as culture, politics, laws, social class, and world events. People and their environments form an inseparable whole; every person's development is affected by all four ecological systems (Bronfenbrenner, 1979/2010).

Thus, instead of copying the Hakka cultural symbols and images, the curriculum designers for the 2021 camp decided to facilitate children's identification of themselves as part of a microsystem, a mesosystem, and a macrosystem of cultural formation. The goal was to convey a comprehensive understanding of the culture and to ensure in-depth knowledge transfer. With respect to the microsystem, we used methods that children can understand to plan intergenerational interactions within families so that family memories and experiences can be transmitted. Meanwhile, regarding the mesosystem, we interviewed and showed videos of Hakka experts growing up and invited children to share their family stories with one another after watching the videos. This approach was designed to encourage children to understand one another's family microsystems as well as the similarities and differences in their own family experiences and those of the Hakka groups. For the macrosystem, we encouraged the children to discuss the challenges faced by Hakka immigrants to gain deeper understanding of their plight and the wisdom of their ancestors. The goal was not only for children to learn about Hakka culture but also to increase their awareness of the relative position of their own ethnic group in a pluralistic society and to be part of their own cultural continuity.

5 Overview of Summer Camp Curriculum

The goal of this 3.5-day summer camp curriculum was for 7- to 12-year-old students to (re)learn about the Hakka diaspora and to consider Hakka cultural artifacts and customs through data visualization.

The course was delivered through an online platform, and we provided a creativity kit to each participant, which included a family interactive worksheet and materials required for creating artwork at the end of each learning unit. These included materials for designing Hakka-style hand-wrapped silk flowers, Hakka tea, and colored yarn for data visualization; in addition, pictures of tiles in various colors were provided to be used to create terrains representing farms of students' Hakka ancestors. Each creativity kit became a gift or treasure box for students to explore; they expressed curiosity about the materials provided and how they would be used. The summer camp was designed for students in Grades 1 through 6, not limited to those with Hakka background.

We designed three units by using the data visualization method.

- Unit 1: Where Did the Hakka Come from and Where Did They Migrate to?
- Unit 2: Choosing Where to Explore
- Unit 3: Comparing the Past and Present

As visualized in Figure 8.1, each unit included preclass preparation, which involved asking the children to use the worksheets to interview their parents and grandparents and open an intergenerational dialog within their family. Thus, they could obtain the basic information required for data visualization. In the formal classes we imparted knowledge of the Hakka culture by inviting "Hakka experts" (i.e., Hakka people who now live and flourish in Taipei) to share their experiences moving to Taipei from their original homes, differences in life and work in both places, and adaptation to their new lives. Next, we guided the children to reflect on the worksheets with which they opened the intergenerational dialog in their families; from there they refined their data. We guided them to use artistic principles, rules, and elements to add their personal feelings and experiences to their data.

5.1 *Unit 1: Where Did the Hakka Come from and Where Did They Migrate To?*

The teaching objective of this unit was to reestablish the children's understanding of the immigration of Hakka clans from China as well as their migration within Taiwan. The students were led to explore the origins of their families through their own ancestral halls and pedigrees. They learned not only how

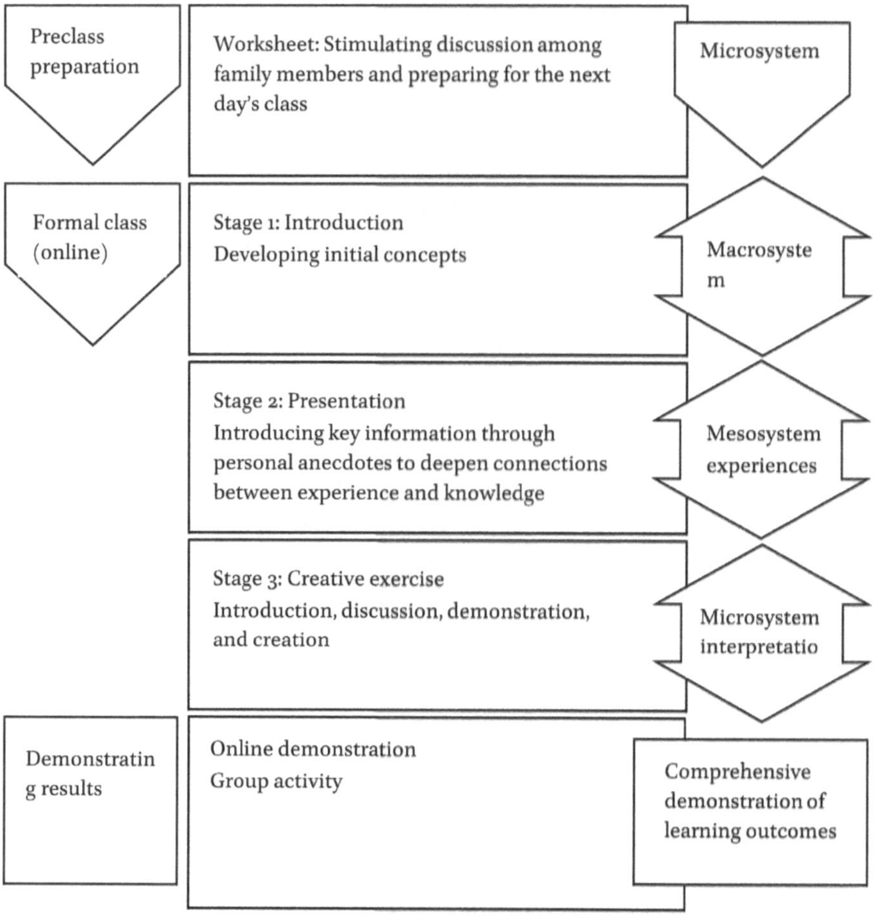

FIGURE 8.1 Teaching method for online summer camp curriculum

Hakka people are distributed across Taiwan but also how those in Taipei migrated from other regions of Taiwan.

In this unit we demonstrated that the locations to which the Hakka people migrated can be used to deduce the time periods during which and reasons for which they migrated. For example, Hakka immigrants moved from Taoyuan and Hsinchu (located close to the northern part of Taiwan) to the eastern and southern parts of Taipei City between 1895 and 1945. As railway and highway transportation increased throughout that period, industry boomed in the Taipei area, providing jobs and entrepreneurial opportunities. Hakka people from all over Taiwan, then, moved to Taipei (Lo, 2020). In other words, the destinations that Hakka families chose reveal their unique heritage and origins. Children who do not have a Hakka background can use the same methods to start a dialog with their families and trace their family history of migration in Taiwan and Taipei while gaining a shared understanding of the migration history of Taipei's second-largest ethnic group.

5.1.1 Unit 1 Content

Before the formal class we asked the children to open their creativity kits and retrieve the worksheet inside. Parents and children were asked to discuss their family, using the designed worksheet as a guide. Parents were invited to share their family's migration experiences, any stories they may know about the locations to which their family migrated, and any memories or sentiments they have about these locations. Parents were also asked to encourage their children to talk to their grandparents about their stories, thus collecting more information for the data visualization tasks.

This lesson, which also marked the official beginning of the unit, introduced knowledge about the migration history of Hakka groups while simultaneously guiding children to discuss the place where their families gather when worshiping their ancestors—the *clan hall*. We discussed clan hall designs, clan hall names, and the meanings behind them, using previously known facts and experiences of students to elicit their unique feelings about their families. We helped the students learn about their own family trees and held an interactive discussion with the children about important threads and elements of heritage within their family trees. We then went into the presentation section, where we played a prerecorded "Hakka Experts" video, in which we interviewed a family of Hakka people who had moved from Taoyuan to Taipei, where they worked their way up from a low economic status to become well-known reformers of traditional markets. They shared the history of their struggle in Taipei. Next, we moved to a creative exercise, in which we led the students to use the stories they previously learned from their parents and grandparents as a starting point to reflect on the contents in the introduction and presentation sections. This also served to connect students' learning from the worksheet to their interactions with their parents and grandparents.

The students prepared for the visualization–creation exercise, using the family migration histories they had gathered via the worksheets. They also calculated the phases and specific years in which their family members moved. In the preceding day's worksheet interaction, the students would have already discussed with their parents the memories and feelings they associated with migration destinations and would also have chosen colors that evoke the feelings each migration destination holds for them. We integrated Hakka hand-wrapped flower techniques into the creative process, guiding the students through two wrapping methods—one traditional and one modern. The colors and number of loops wrapped onto the flower stems were to represent the family's movements within Taiwan. In other words, from the color and loops of silk wrapped on a student's flowers, we could visually understand the migration history of the student's family.

5.1.2 Student Work and Key Takeaways

Prior to the class the children had only a vague understanding of Hakka culture. Through the in-class activities they developed a deeper understanding of their own family histories and migration experiences. This was the first time they had attempted to create art using data visualization. Most of the children found converting the anecdotal migration history into raw data and then transforming the data into artistic renditions challenging. They did not know how to use the data as a basis for artistic creation. We asked the children to use the arithmetic method of addition to calculate how many years their families resided in each place. Next, we invited the children to work with their parents or grandparents to choose a color for each place their family had resided, advising them to use color to represent their feelings for the place. We then linked the preparations noted above to the creation of Hakka hand-wrapped silk flowers. The physical metaphors of the wrapped flowers included the number of looped turns of silk thread (symbolizing the number of years the families had stayed at each location), and the colors that were associated with each location through the students' family discussions. Once the children were able to use the provided materials to make the wrapped flowers, they gained a better understanding of the method and learned that data visualization relies heavily on raw data collection and preparation. We present some of the participants' realizations below.

Participant L (11-year-old student): "Our family moved so many times, my grandparents don't even know how many, but they still remember several major moves they undertook. I used different colors to express their feelings and memories" (Figure 8.2).

Participant K (10-year-old student): "My maternal grandma said the home she liked the most was the house on the Niulan River, Longtan [in Taoyuan County]. But the area developed, so they had to build an overpass there. Our family's land was confiscated, so they were forced to move from there. Grandma said a clear river ran there; she really misses the simple, heartwarming times from when she was little, catching fish and crayfish and washing clothes. Those

FIGURE 8.2 Left to right: participant L's and K's Data Visualization Project from Unit 1

feelings are very warm for her, so we chose orange thread to represent grandma's feelings about that home" (Figure 8.2).

Unit 1 lesson, therefore, integrated the use of the worksheet, the introduction of objective knowledge like the migration history of Hakka groups, data collection, and artistic–visual analogies. In this lesson children engaged in a planned, systematic artistic creation that embodied their family memories and heritage. These constituted the major goals of the unit's data visualization.

5.2 *Unit 2: Choosing Where to Explore*

The teaching objectives of Unit 2 were to use shape and color as the basis for terrain identification and to use data visualization to facilitate an understanding of the geographic choices made by Hakka people as they migrated from China to Taiwan and the manner in which they adapted their farming practices to Taiwan's local conditions. The Hakka people migrated to Taiwan for a variety of reasons, but in many cases the first destination they chose determined the kinds of activities the clan could pursue. Our creativity kit for this unit contained precut shapes that could represent terrains like hills or basins. Students could use the materials to show how their ancestors adapted to cope with new environmental challenges. We asked the students to explore how environments, represented by the colored terrain pieces provided in the creativity kit, could present challenges. After the initial migration of the Hakka people to Taiwan, they settled in areas with a range of geographies and landscapes, including mountains, hills, and basins, each presenting its own natural resources, such as water sources, trees, and arable land. The students learned about the geography of the settled areas, including their height, slope, watersheds, arable areas, and the conditions of life there. Step by step, the children layered the colored pieces from our creativity kits to form hills and plains. They were required to use the knowledge they gained to think what produce could be grown on the terrain they created.

5.2.1 Unit 2 Content

Before the class began, students' parents were invited to give a presentation on the geography and environment of the place in which they and their parents (that is, the children's grandparents) grew up, any agricultural products that may have been grown on their land or on the land close to where they lived, and any childhood experience of working on the farm.

Before the introduction, we invoked a Hakka saying: "Wherever there's a mountain, there's a Hakka." We asked the students about the location of their ancestral homes in Taiwan and what made the land of each location special as well as the agricultural products of that region. Next, we introduced the two schools of thought regarding Hakka migration to higher elevations. Traditional academics have consistently stated that Hakka people were forced to live in hilly areas and mountain foothills because the plains were already occupied by the Hoklo when the Hakkas arrived in Taiwan. According to the hypotheses of other historians, however, early Hakka migrants voluntarily moved from the plains to higher elevations (Chang, 1999). We then asked the students to discuss which theory they supported and their reasons for doing so. In the presentation section we invited a Hakka elder who runs a tea business in Taipei and who has achieved the rank of tea master to introduce the students to the relationship between Hakka settlements and tea through a video. The tea plant (Camellia sinensis) grows well on hilly slopes, and hilly regions are precisely the areas to which Taiwan's Hakka people migrated; consequently, tea was a major crop for Hakka farmers. Thus, tea became an important symbol in Hakka culture. The tea master introduced the regions of Taiwan where tea is grown, the tea leaf drying process, and a signature Hakka variety of tea—Oriental Beauty Tea. She also introduced various techniques for tasting tea. The children were then asked to practice tea tasting at home, using the tea provided in the creativity kit. In the creative exercise section, all students received a "terrain bag" in their creativity kits, which included a puzzle representing various terrains. Each student was provided with a unique terrain. A note was included informing students that they were to pretend to be in charge of part of the terrain and to build their own farmland and homes in accordance with the features of the terrain.

5.2.2 Student Work and Key Takeaways

In Unit 2 we asked the children to imagine themselves as landowners with large plots of land so they could plan the agricultural products they wished to grow. We included colored sponge sheets in the creativity kits, which the students were then asked to cut into semicircles of various sizes and then group or stack the shapes to represent the topography of the farmland; dried flowers were used to represent cash crops. The children were then able to "plant" the crops on their "land." In this unit children were asked to create their own piece of land using the materials provided; some children even stated that they imagined themselves as real landowners during the project. We present some of the participants' description of their lands below.

Participant B (8-year-old student): "I was given some land that was white and kinda low hills. I grew trees. Although the white part was my primary planting area, I couldn't stop myself from planting on the yellow part, too. I wanted to plant fruit trees."

Participant C (8-year-old student): If the character in my design gets tired from walking, he can lie down at the edge of the field and enjoy the scenery" (Figure 8.3).

Participant D (9-year-old student): "My land is on a single hill. I want to grow tea because the Hakka people grow tea in the hills. But I also want to add a slide in the tea garden! The hill is perfect for making a slide! I want to plant tea and have fun at the same time!" (Figure 8.3).

FIGURE 8.3 Left to right: projects of Participant C and D from Unit 2

Choosing Where to Explore is a lesson on the Hakka people living in harmony with their natural environments. By understanding the migration history and cultivation practices of their own families, we hoped to help students form deeper connections with the land, gain a better understanding of the challenges and opportunities that each terrain posed, and learn how people can coexist with others.

5.3 *Unit 3: Comparing Past and Present*

In the design for Unit 3, we used a representative Hakkanese phrase, "晴耕雨讀 *ciang gang, rhi tug*" ("Farm on sunny days. Study on rainy days") to represent the Hakka notion of self-development. The phrase is so prevalent that many families use it as their family motto; indeed, it is often carved into the plaques displayed on the lintels in ancestral halls (a position of great reverence). The phrase reflects the extent to which Hakka people cherish both time and their desire for learning.

The teaching objectives of Unit 3 were twofold. First, to encourage the children to compare their lives with their parents and grandparents' childhoods, in this unit we aimed to promote an understanding of the changes that occurred in the living environment. The comparisons the children drew served as a form of connection, helping them to view family members in a unified family history and legacy. Second, based on the comparison of the experiences of the three generations, the children performed data calculations and used data visualization to present how the lives of Hakka people changed over generations.

5.3.1 Unit 3 Content

In the preclass preparation, parents, grandparents, and children were asked to hold a family discussion using the provided worksheet as a guide. We asked the students to fill in the worksheet with both their answers and those of their grandparents. We also encouraged them to ask more questions that were not listed on the worksheet and to collect more information that could be used to make comparisons. Six questions were included on the worksheet:

- Q1: Grandpa/Grandma, what did you eat for meals when you were a child?
- Q2: Grandpa/Grandma, when could you buy new clothes when you were a child?
- Q3: Grandpa/Grandma, what kind of house did you live in when you were a child? What was it made of? How many floors were there?
- Q4: Grandpa/Grandma, how many vehicles (cars, motorcycles, motor scooters) did you have when you were a child?

- Q5: Grandpa/Grandma, when you were a child, what kinds of toys did you have? How much time were you given to play with them? What kinds of games did you play, and how long were you allowed to play them?
- Q6: Grandpa/Grandma, when you were a child, how long did you spend studying every day?

In the formal introduction section we asked the children to discuss the Hakka saying "Farm on sunny days. Study on rainy days." They shared their understanding of where the saying originated and compared the rural lives described in the saying and the urban environments in which they now live. We presented them with photos of past and present living environments and asked the children to identify differences in environment and climate.

For the presentation section we created a video of an "agricultural expert," featuring 80-year-old Mr. Liu and his son sharing stories about farming, the tools they used, and the scale of tasks they had to complete each day. Mr. Liu also shared how farming shifted from animal-assisted to mechanized. His son, by contrast, shared how he changed careers from young farmer to civil servant and how his life changed accordingly. The story of this Hakka father and son reflected Taiwan's shift from an agricultural economy to an industrialized one.

In the creative exercise section, based on Mr. Liu's and his son's stories, we first asked the students to review the information they collected the day before from their parents and grandparents. To improve their understanding of how data can be expressed creatively, we taught them the concept of infographics and introduced them to *Dear Data* (Lupi & Posavec, 2016), an art project by two British information designers, Georgia Lupi and Stephanie Posavec, who sent each other a data visualization postcard every week for a year, using colors, lines, and patterns to visualize and share data from their daily lives with each other. After considering the content of *Dear Data*, students were asked to refer to the information from their worksheets and visually share their stories and the information they learned the previous day. We hoped students would successfully draw inferences about their own lives and their grandparents' childhoods. The children then used colored pens, tape, stickers, colored paper, and other provided art materials to present their results from the calculations and to make comparisons across generational experiences.

Comparisons revealed differences in living arrangements. For example, when one student asked his grandparents, "Which floor did you live on when you were a child?" the grandparents answered, "The only floor—the first." The majority of original Hakka village dwellings were traditional single-story bungalows. This student, by contrast, lived on the eighth floor of an apartment building with extended family members living on the other floors. Thus, the floors on which generations of the same family resided demonstrate a marked contrast in lifestyles. Some students asked their grandparents how long they kept the lights on at night when they were younger. The grandparents replied that all lights were out by 8 p.m. The children, however, noted that they themselves had to study until 10 p.m. Thus, the children identified the amount of lighting to which the two generations had access. Visualizing data makes differences between generations more prominent than when they are described with text or numbers.

5.3.2 Student Work and Key Takeaways

The master farmers in the video were a father and a son, the father's age corresponding to that of many of the children's great-grandfathers. The video was primarily in Hakka, which, most of the children admitted, they would not be able to understand in the absence of subtitles (or assistance from adults), highlighting the loss of Hakka linguistic heritage. Mr. Liu was much older than the children's grandparents, so even their grandparents' experiences differed from his, providing an additional point of comparison. Mr. Liu's and his son's experiences further helped the children understand past and present agricultural lifestyles, what farming methods still apply, which methods evolved, and which methods no longer exist. Unexpected comparisons of their own experiences and their grandparents' childhoods surprised the children.

For example, Participant D (9-year-old student) stated: "My grandparents lived in a bungalow when they were kids, but now I live on the fifth floor of a building. My grandparents only got new clothes on Lunar New Year when they were kids, but I get new clothes at least six times a year." As shown in Figure 7, Participant D visualized the data—numbers and types of clothes—by placing one pink tape next to six pink tapes. Participant D added: "Grandpa raised dogs, cats, cows, pigs, chickens, and geese when he was a child, but I only have one dog." She represented the animals with six round stickers embellished with hand-drawn animals placed above a single sticker of an image of dog. She first used six different-colored round stickers to represent the six animals that her grandfather raised when he was a child. Next, she shaped the stickers to represent the characteristics of the animals. For example, she cut out a bitemark on the first silver sticker from the right in the row of six to show that the dog her grandfather raised as a child was very ferocious unlike her own pet dog (she added a cute dog sticker to the other silver sticker). The fourth and fifth stickers from the right in the first row were both gold, but she distinguished them with cutouts to represent chickens and geese (Figure 8.4).

In another example, Participant J (10-year-old student) stated: "Grandpa told me that when he was a child, he ate sweet potatoes for breakfast, lunch,

FIGURE 8.4 Left to right: Participant J's and D's data visualization project from Unit 3

and dinner! He told me he had only two snacks per day, but the snacks were also sweet potatoes. His choices for side dishes were mostly vegetables and pickled squash. This surprised me so much! I can eat Japanese food, spaghetti, and hot pot. Poor Grandpa! Life in the past was way harder than it is now!" As shown in Figure 8, Participant J used different colors to represent types of food. For example, on the lower left side, one piece of the visual data shows a red square representing hot pot, a green square representing Japanese food, and two wavy-lined squares representing spaghetti. In the upper section, the visual data show multiple orange squares representing sweet potatoes, a black square representing pickled squash, and several yellow squares representing other vegetables. Using a variety of colors and numbers of squares, Participant J visualized the foodways of her grandfather as well as her own.

Participant J also visualized her and her grandmother's daily routines and emotions associated with those routines. For example, as shown in the upper right side of Figure 8, their sleeping and morning routines were represented by the dark-colored line, with Participant J on the left and her grandmother on the right. Participant J's grandmother said she enjoyed going to school and doing her homework; accordingly, Participant J drew a lot of hearts and smiles on the white paper tape representing the school hours. Participant J did not enjoy school, so her school hours were represented by a wide dotted white tape with no emoticons. She hated after-school classes so much that she drew angry faces all over another white tape representing after-school hours. Playing with siblings after school was Participant J's grandmother's happy time; dinner was the happiest time for J, so she used a gold sticker to indicate these favorable emotions and events. Based on the lengths of the line and tape, colors, and emoticons, Participant J visually recorded her and her grandmother's daily routines (Figure 8.4).

6 Conclusion

Following the creative data visualization approach, children experienced three stages of learning. The first stage was research. Each student was expected to conduct research (appropriate to their age level) regarding their own family history, to better understand their own origins and those of their families, clans, and Taiwan. Without this stage, the creative use of data would be impossible because such exploration is key to promoting an in-depth understanding of their roots and Hakka culture. The second stage included data compilation, in which the children were expected to extract and compile data from the information they collected. Other elements, such as color and choice of materials, were added to symbolize associated imagery. The children engaged metacognitively as they considered how to interact with the physical materials, colors, and symbols to systematically code raw data into a visual narrative. The third stage was data visualization and presentation. At this stage the students added their own interpretations to their stories and transformed objective raw data into subjective visualizations that better connected information for the viewer. Once viewers understand the logic behind the data visualization, they can interpret the message with minimum error, facilitating communication across linguistic boundaries and cultural heritages.

The summer camp organized by the Taipei Hakka Culture Foundation applied data visualization and family discussions to facilitate education on Hakka culture. Instead of obtaining knowledge from textbooks, the children were encouraged to construct knowledge from their own family histories. In addition, other Hakka cultural elements, such as sayings, cultural artifacts, family histories, and traditional craftsmanship, were introduced. Overall, the online camp established the following core concepts of cultural literacy:

- connecting closely with family,
- understanding of ethnic groups more deeply,
- learning to compile information objectively, and
- learning to interpret stories based on and through raw data.

Foundation leaders considered this learning model to hold potential long-lasting effects for cultural education because the knowledge becomes internalized. The skills students acquired helped to deepen their understanding of Hakka culture as well as their abilities to respond to more general challenges they may face in the future. The goal of this curriculum was to ensure that Hakka culture is not reduced to mere symbols and to use a learning process that develops competencies engaging with the micro-, meso-, and macrosystems of the culture.

Humanity has now entered the data age, and all students must develop data literacy—the skills necessary to read, work with, analyze, and argue with data. "Reading" means to recognizing what data are and what aspects of the world those data represent. "Working with" means having the skills to create, access, and organize data. "Analyzing" means filtering, ordering, organizing, and comparing data. Finally, "arguing with data" means using data to communicate with and persuade an audience (Bhargava & D'Ignazio, 2015). In our case study data visualization was demonstrated through creative works and pedagogical strategy, including

- quantifying of experiences to turn them into visual products,
- transforming data into a line of reasoning that can convey information, and
- demonstrating the data through collation and comparison of experiences.

Whether a data visualization teaching plan comes to successful fruition depends on whether the teacher has guided the students toward preliminary data collection, that is, thematic research. This sets the present data visualization-based pedagogy apart from art classes of the past, which relied on children's spur-of-the-moment improvisational creativity. Considering the importance of data literacy, data visualization provides a way for students to build skills in reading, working with, analyzing, and arguing with data (that is, skill in data literacy). Drawing on personal experiences and explicating them through colors, threads, and textures can transform objective, hard data into delicate, emotive expression. Data visualization as we have employed it in this lesson plan adds the much needed crucial dose of understanding to data literacy.

References

Bertling, J. G., Hodge, L., & King, S. (2021). The case for data visualization in the art classroom. *Art Education, 74*(2), 44–49. https://doi.org/10.1080/00043125.2020.1852381

Bhargava, R., & D'Ignazio, C. (2015). Designing tools and activities for data literacy learners. *Web science: Data literacy workshop*. http://www.kanarinka.com/wp-content/uploads/2015/07/DataLiteracyRahulCatherine.pdf

Bronfenbrenner, U. (2010). *The ecology of human development: Experiments by nature and design* (Liu-Chen Zeng, Trans.). Hsin Li Press. (Original work published 1979)

Chang, C. J. (1999). A page of vicissitudes of life: 400 years of Hakka people in Taiwan. *Taiwan panorama*. https://www.taiwan-panorama.com/Periodical/Details?Guid=3f36e5f4-6b99-4ef6-8215-34cdf4389733&fromYear=1999

Chiou, C. T. (2012). The past and present of Taiwanese Hakka people. *Journal of Lifelong Learning of Taipei City, 57*, 2–12.

Cooper, Y. (2014). The creation of data visualization and synthesizing process. *Wikicreation*. https://www.academia.edu/9703528/The_Creation_and_Data_Visualization

Cooper, Y. (2017). Looking at American STEAM education from the A of STEAM. *Hong Kong Arts Education Journal, 2017*(1), 4–9.

D'Ignazio, C., & Bhargava, R. (2018). Creative data literacy: A constructionist approach to teaching information visualization. *Digital Humanities Quarterly, 12*(4). http://www.digitalhumanities.org/dhq/vol/12/4/000403/000403.html

Huang, Y. C. (2012). *The "invisibilization" of urban Hakka: The experience of ethnic identity* [Unpublished master's thesis]. National Central University, Taiwan.

Letouzé, E., Noonan, A., Bhargava, R., Deahl, E., Sangokoya, D., & Shoup, N. (2015). *Beyond data literacy: Reinventing community engagement and empowerment in the age of data*. Data-Pop Alliance White Paper Series. http://datapopalliance.org/wp-content/uploads/2015/11/Beyond-Data-Literacy-2015.pdf

Liao, C. (2019). Creating a STEAM map: A content analysis of visual art practices in STEAM education. In M. Khine & S. Areepattamannil (Eds.), *STEAM education: Theory and practice* (pp. 37–55). Springer. https://doi.org/10.1007/978-3-030-04003-1_3

Lo, L. S. (2020). *Survey and research project plan on Hakka culture in Taipei city*. (unpublished). Taipei Hakka Culture Foundation.

Lupi, G., & Posavec, S. (2016). *Dear data*. Chronicle Books.

National Coalition for Core Arts Standards. (2014). *National core arts standards: A conceptual framework for arts learning*. https://www.nationalartsstandards.org/sites/default/files/NCCAS%20%20Conceptual%20Framework_0.pdf

Urist, J. (2015, May 14). From paint to pixels. *The Atlantic*. https://www.theatlantic.com/entertainment/archive/2015/05/the-rise-of-the-data-artist/392399/

Yang, W. (2004). *National Hakka population basic data survey and research*. Commissioned by the Executive Yuan Guest Committee. http://ir.sinica.edu.tw/handle/201000000A/66630

CHAPTER 9

Promoting DEI with STEAM Technology for People with Visual Impairment and Blindness in Taiwan

Hsin-Yi Chao

Abstract

Essential spiritual nourishment is crucial for one's psychological well-being. Art aesthetics serve not only as a medium for expressing personal emotions and thoughts but also as a channel for communication and dialogue between individuals and their social environments. Furthermore, art aesthetics are foundational in personality development and nurturing. This chapter examines the future directions of visual arts education for people with visual impairments and blindness in Taiwan and explores ways to foster a social environment conducive to inclusive learning.

Keywords

visual impairment – blindness – DEI – art education – STEAM

1 Introduction

Cross-disciplinary learning for individuals with disabilities benefits from the integration of technology and engineering applications, particularly in educational settings like schools and museums. This approach addresses challenges such as limited comprehension of perceptual information or the need for alternative content interpretation (Bandyopadhyay & Rathod, 2017; Hasper et al., 2015; Lahav et al., 2019). By incorporating STEAM (Science, Technology, Engineering, Arts, and Mathematics), accessible technological designs can help visually impaired individuals better understand art and scientific concepts. This approach leverages STEAM's holistic characteristics to education to promote inclusive learning experiences across disciplines for individuals with disabilities. Based on practices in Taiwan, this chapter will discuss teaching strategies, the development of teaching materials, technology, and STEAM

 | DOI:10.1163/9789004719200_010

applications that guides museum autonomy for individuals with visual impairments and blindness.

2 Current Status of Education for People with Visual Impairment in Taiwan

In Taiwan, the definition of visual impairment has evolved considerably over time. Initially, the definition was influenced by individual social networks prior to Japanese rule and based on perceived capabilities and characteristics. During the early period of Japanese governance, assessments were carried out by medical professionals. Since the 1980s, following the implementation of social welfare programs, individuals with visual impairments have been evaluated according to a variety of instrumental vision measurements and classification standards. These standards are now part of a dynamic approach by the state, and encompass comprehensive national definitions, assessments, legal frameworks, knowledge, treatments, and assistive technologies (Chiu, 2013).

According to a report by the World Blind Union (2014), which is an international non-governmental organization, approximately 285 million people worldwide are living with visual impairment. The current report from the Ministry of Health and Welfare (2023) records 1,214,668 people with disabilities in Taiwan, which comprises approximately 5% of the total population, with 54,155 registered as visually impaired. The significant impact of visual impairment on daily life highlights the crucial need for improving people's everyday experiences, which is a pressing societal issue.

The statistics of The Gender Equality Committee of the Executive Yuan (2023) on the number of students with physical and mental disabilities in special education at all school levels record 117,185 students with disabilities in tertiary institutions and below. Of these, approximately 1,231 are visually impaired (710 males and 521 females). Of these students, 670 are in high school and below, while 561 are enrolled in tertiary institutions. Educational institutions include specialized and general schools for concentrated special and inclusive education, respectively, with various class types such as concentrated special education classes, dispersed resource classes, itinerant counseling, and regular classes offering special education services.

2.1 *Current Status of Special Schools for People with Visual Impairment*

The history of education for people who are blind in Taiwan can be traced back to 1891 when Rev. William Campbell, a Scottish Presbyterian missionary, founded the "Training Hall for the Blind" (Chiu, 2013). Pioneering in its approach, this institution began teaching people who are blind braille, arithmetic, and handicraft, thus marking the inception of specialized education for people with visual impairment in Taiwan. Today, Taiwan hosts three primary special schools dedicated to students with visual impairments: the Taipei School for the Visually Impaired, the Taichung School for the Visually Impaired, and the Huei-Ming School for Blind Children. Additionally, there are 21 special schools with all kind of disable students in different cities and counties including the visual impaired (Ministry of Education, 2023).

However, these specialized institutions often face challenges due to a lack of faculty specializing in the arts and humanities. The focus within their arts programs predominantly leans toward music and auditory training. Moreover, the availability of fine art materials in braille is limited, which complicates the delivery of fine art classes. Consequently, these schools frequently depend on external educators and additional resources to enhance their fine art education offerings.

2.2 *Current Status of Inclusive Education for People with Visual Impairment*

Inclusive education for people with visual impairment emphasizes integrating students into standard educational settings to foster a non-segregated environment, which prepares them for social interactions from an early age. The learning environment in inclusive settings is rich in visual information and spans various subjects such as languages, mathematics, natural sciences, social studies, arts, and humanities. Such information includes texts, numbers, tables, images, photos, and teaching aids. Students with visual impairment must access this content through tactile means, such as braille or tactile graphics.

Particularly in the fine arts, understanding and manipulation rely heavily on tactile skills. Educators must adapt the pace of teaching to meet the abilities of students with visual impairment and provide assistive devices and concrete learning experiences. Strategies such as simplifying content, reducing information volume, reanalyzing, and reorganizing, or employing alternative sensory materials as learning media are essential to enhance learning outcomes for these students (Axel & Levent, 2003).

2.3 *Challenges in Fine Arts Education for People with Visual Impairment*

In Taiwan's special schools for people with visual impairment, art programs have predominantly focused on music education, often to the detriment of visual arts. Recent efforts to incorporate external educators for fine arts-creation courses or to utilize resources from museums and art galleries for art appreciation activities have faced challenges in terms of establishing a standard learning environment for students with visual impairment.

Key challenges within inclusive education include training mainstream teachers to effectively guide visually impaired students in fine arts courses and to develop multisensory materials to improve their understanding. The heavy reliance on teacher assistants to help complete art projects is a significant ongoing challenge, which underscores the need for further innovation and support in this area (Metatla, 2017).

3 Development of Fine Arts Education for People with Visual Impairment in Taiwan

Fine arts education should be a universally accessible resource for all school-aged students, encompassing skills in language and cognitive expression, manual dexterity, spatial orientation, tactile learning, and the development of social relationships and personal integrity. Moreover, such education should foster life skills necessary for independence, thereby establishing a foundation for lifelong learning (Axel & Levent, 2003).

Traditionally, Taiwanese society and educational institutions have overlooked the artistic educational opportunities available to students with visual impairment. This neglect is highlighted by a notable scarcity of visual arts courses and a lack of museum resources for visitors with visual impairment. Few private organizations have offered fine arts activities, primarily due to prevailing ableism, which underestimated the need for art education among people with visual impairment. This was further compounded by a lack of qualified teachers and professional expertise necessary to create a comprehensive fine arts education environment for this demographic. In recent years, with the government's promotion of cultural equity policies and increased advocacy by civil society, the availability of artistic resources for people with visual impairment has witnessed gradual improvement.

3.1 *Nonprofit Organizations for Fine Arts Education of People with Visual Impairment*

In Taiwan, over 100 nonprofit organizations, including foundations and associations at both national and regional levels, primarily serve individuals with visual impairment. These include the Blind Welfare Association across various counties and cities, Parents Associations for the Visually Impaired, and Care Associations for the Visually Impaired, among others. These groups provide services across a wide range of age groups and social demographics, including care homes and rehabilitation centers for people who are blind, focusing on primary educational subjects, social welfare advocacy, self-care rehabilitation, and vocational training.

In 2014, Taiwan established the Taiwan Art Beyond Vision Association, the sole public welfare entity dedicated to providing fine arts education services for people who are visually impaired. This organization continues to offer fine arts courses for students and adults, initiate non-visual aesthetics lectures, and develop multisensory teaching materials and methods.

3.2 *Fine Arts Education for People with Visual Impairment in Special and Inclusive Education*

Fine arts courses initially intended for sighted students can be adopted for people with visual impairments by incorporating tactile guidance alongside step-by-step verbal instructions. This approach allows students with visual impairments to actively participate in creating art using different mediums. For students with low vision who can utilize residual sight for color identification and spatial orientation, visual guidance and tactile manipulation training are incorporated. For students who are completely blind, courses focus primarily on full tactile guidance and detailed verbal descriptions of images, converting elements such as color, lines, and shapes into a comprehensive learning experience of tactile and auditory cognition.

Consequently, six capabilities for learning fine arts as an individual with visual impairment have been developed, thereby serving as a foundation for creating fine arts educational materials for people with visual impairment and providing a reference for teachers in inclusive education settings (Figure 9.1):

- Manual Tactile Manipulation Capability: Fine arts creation activities for people with visual impairment not only require learning how to handle tools through tactile manipulation but also require a more refined ability to discern the surface textures of objects or to judge the spatial environment of the creation activity through touch. Students with partial vision are encouraged to enhance their understanding of art creation by using touch, smell, or hearing, thereby leveraging these senses to compensate for visual limitations.
- Image Recognition and Communication Capability: This involves learning how visual images represent three-dimensional objects as two-dimensional images. It aims to build an internal "image database" in the mind of people with visual impairment, which can be accessed to link tactile perceptions with stored visual memories. This capability facilitates communication using visual symbols and enhances decision-making by enabling students with visual impairment to share common visual understanding with sighted peers.

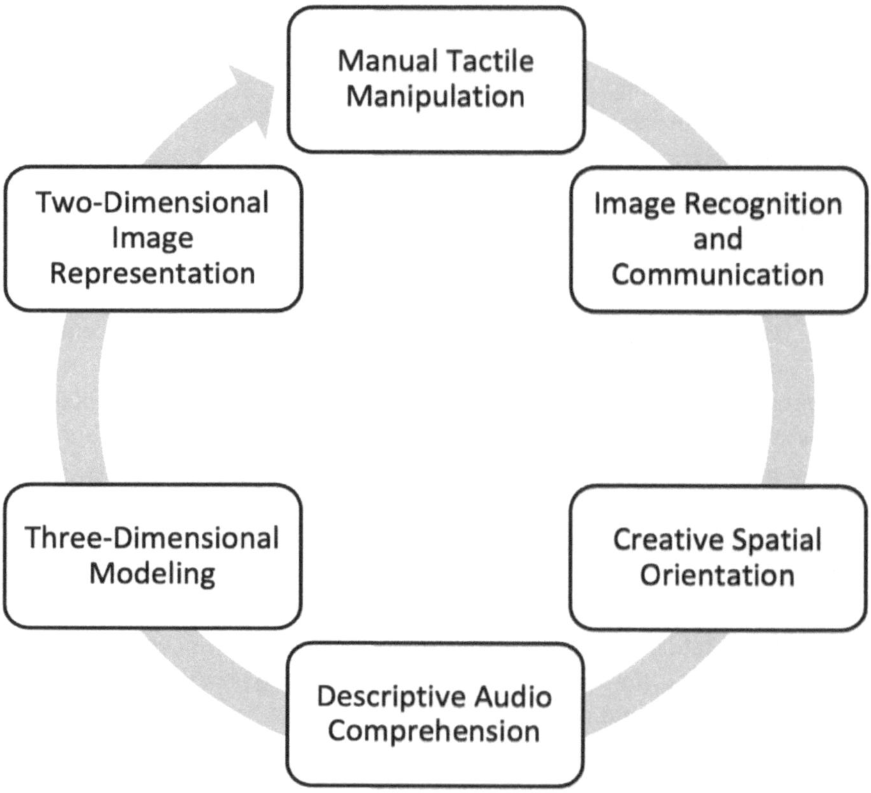

FIGURE 9.1 Six capabilities of visual art education for students with visual impairments and blindness

- Creative Spatial Orientation Capability: Essential for both mobility and art creation, this capability involves understanding the relationships between body movement and space. It connects cognitive processes that assess direction and distance, which is crucial for arranging elements in art projects. In practical terms, it means using tactile feedback from finger movements or tools to navigate and arrange artistic components effectively.
- Descriptive Audio Comprehension Capability: When engaged in art creation, students who are completely blind rely heavily on verbal description provided by instructors, parents, or volunteers. This information, which includes details about color, length, orientation, size, and shape, must be communicated precisely and succinctly. Students with visual impairment must then accurately interpret these descriptions and translate them into tactile or other sensory responses that aid in their artistic expression.

- Three-Dimensional Modeling Capability: This capability focuses on creating art that involves actual spatial dimensions, which can be more intuitively understood by students with visual impairment through touch rather than flat images. This is developed through direct interaction with various materials such as clay, thereby enabling students to explore and express different textures and forms. Starting with basic elements like points, lines, and planes, and then applying them in complex ways, it also involves scaling from small to large three-dimensional models to further connect with bodily kinesthetic senses.
- Two-Dimensional Image Representation Capability: Addressing the limitations in existing tools for artists with visual impairment, this capability involves the use of alternative media to represent flat images. Techniques such as raised lines or tactile depressions created on soft materials help convey the structure of flat images. This method allows students with visual impairment to perceive and understand the planar arrangement of elements such as points, lines, and planes, thus enhancing their grasp of spatial concepts and perspective.

3.3 *Museum Accessible Technology and Cultural Equality for People with Visual Impairments and Blindness*

The Metropolitan Museum of Art in the United States initiated tactile art recognition activities in 1913, which marks one of the earliest records of museum services tailored for people with visual impairment (AAM, 1992). Subsequently, many countries have endeavored to develop accessible services and multisensory exhibition designs. In Taiwan, in 1993, the Taipei Fine Arts Museum began providing tactile guide services for people with visual impairment, and these services have gradually expanded across various museums and art galleries. These include themed exhibitions designed with tactile and auditory guide resources to complement the visual displays. In 2013, to further embed cultural equity, the Ministry of Culture designated the National Taiwan Museum of Fine Arts (NTMoFA) as a "model institution" for services to people with visual impairment (Chen & Chang, 2013).

In terms of accessible technology applications, in 2016, the NTMoFA launched the first integrated navigation resource for visitors who are visually and hearing-impaired, the "National Taiwan Museum of Fine Arts Accessible Application". This app encompasses a range of guides on themes such as sculpture (Chao, 2018), photography (Chao, 2019), painting, architecture (Chao, 2020), and mixed media. This APP is supported by iBeacon indoor navigation technology, descriptive audio, sign language videos, tactile relief maps, touchable exhibits and educational materials, tactile guide devices, and braille and

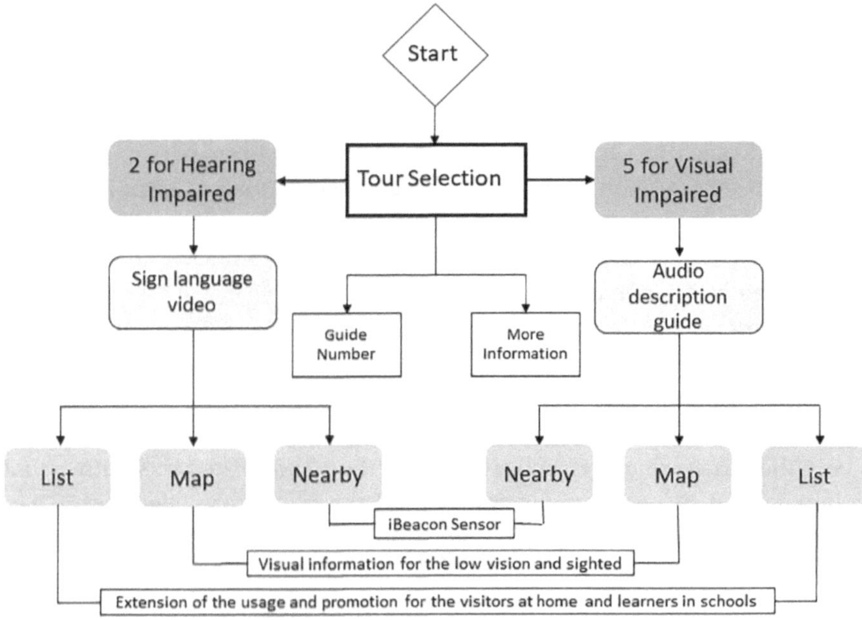

FIGURE 9.2 Functional flow chart of National Taiwan Museum of Fine Arts Accessible Application for the visual and hearing-impaired audience features three guide models: nearby, map, and list with extended usage. Created by Hsin-Yi Chao

enlarged print information, thereby enabling visually and hearing-impaired visitors to conduct self-guided tours with three guide models, nearby, map, and list for multiple functions (Figure 9.2). This technology not only allows students and adults who are visually impaired to transcend traditional group visit limitations but has also been incorporated into special school curriculums as art education materials, thus exemplifying the best in museum-school collaborations for arts resources for people who are visually impaired (Su et al, 2018).

In addition, drawing on both international and Taiwanese experiences with museum services for the visually impaired, Chao (2016) has compiled a table analyzing exhibition space planning, visitor services and assistive tools, multimedia resources, and extended activities as reference for professionals (Table 9.1).

Recently, other museums in Taiwan have also begun to emphasize resources for visitors who are visually impaired, including the National Palace Museum, the National Taiwan Museum, the National Museum of Taiwan History, the National Museum of Prehistory, and the National Center for Traditional Arts. Themes such as autonomous guide technology development, training of professionals for descriptive audio, production of tactile teaching materials and

TABLE 9.1 The methods of visual art exhibition designed for the blind

Subject	Item	Content
Plan of exhibition space	Visiting route	One-way visiting route. Guide path is larger than 40 centimeters in width. Handrail of visiting route with different materials. Audio alarm at visiting corner. Micro-orientation or GPS guide installation.
	Environment	Adjust illumination to the artworks. Increase the contrast of colors of walls and ground.
	Display	360 degrees rotatable display booth. Display bracket 75 centimeters in height. Distance between two works is larger than 140 centimeters. Labels with light color text and dark background.
	Touch Gallery	Permanent exhibition with tactile display. Touchable 3D map with Braille and auditory information.
Visitor services and aids	Space information	3D model of museum architecture Visual description of orientation and mobility.
	Aid and materials	Braille and large-print guide textbook. Tactile pictures of artworks with braille and large print. Optical aids such as magnifiers, projectors, myopia glasses, reading glasses.
	Services	Visiting company. Guide dog company. Guide tour with audio description Touch tour.
	Artworks	Touchable original artworks with gloves. Reproduction as the same texture and size as the original. Raised-line painting. Layered tactile pictures. Replicated collection.

(*cont.*)

TABLE 9.1 The methods of visual art exhibition designed for the blind (*cont.*)

Subject	Item	Content
Multimedia resources	Technical aid	Braille computer. Audio color sensor. Online learning resources.
Extending resources and activities	Workshop	Multi-sensory workshop. Training for art teachers and volunteers.
	Artistic lesson	Lectures of art appreciation, art history, and art criticism.
	Schools and nonprofits	Exchange assistive materials of artworks with schools and nonprofits.

exhibits, training of seed teachers for creative workshops, and accessible exhibition design are the current focuses of museum professionals. These efforts aim to provide relevant artistic resources to special schools or inclusive education settings for students with visual impairment, thereby extending the depth and breadth of visual arts education from schools to society and enhancing career opportunities in the arts for people with visual impairment.

In line with the new museum definitions emphasizing accessibility and inclusiveness, the following trends can be summarized for museum services aimed at audiences with visual impairment:

- Transitioning from specialized tactile visiting areas to incorporating touchable exhibits and braille dual-vision descriptions throughout all exhibition spaces.
- Shifting from single-material tactile designs to a variety of material surfaces that enhance tactile experiences.
- For regularized visitor services, future developments will leverage innovative spatial location technologies to move from guided tours to autonomous visiting formats.
- International museums are increasingly involving individuals with visual impairment in the planning and service design process, thereby ensuring exhibitions are more aligned with the principles of cultural proximity.
- Museums and art galleries are motivating visitors with visual impairment to participate in art creation experiences, thereby elevating the psychological interaction levels between audiences with visual impairment and the artworks.

4 Practice and Development of DEI Principles

Diversity, equity, and inclusion (DEI) principles as well as STEAM applications are crucial in the development of an accessible and inclusive fine arts education. For people with visual impairment, fine arts education should include multisensory operational and cognitive methods. A need exists for a broader understanding of descriptive imagery to provide more visual information to accumulate visual concepts for people with visual impairment, innovative technologies to solve the problem of the non-comprehensibility of flattened images, and more accessible environments to safely and comfortably involve people who are visually impaired in artistic and cultural activities. Particularly, visual arts creation not only enriches the aesthetic education experience akin to that of individuals who are fully sighted but also indirectly trains people who are visually impaired in mathematical and logical thinking, as well as in skills for daily living.

When art education overly emphasizes visual information, it often neglects the right to interact through other senses. By utilizing multisensory stimulation, the interaction in art appreciation can become more comprehensive and complete. "Beyond vision" does not imply the unimportance of vision but attempts to break the confines of visual perception in understanding art, thus emphasizing the use of auditory, tactile, gustatory, olfactory, and kinesthetic senses (Chiu, 2016). Through the art creation and appreciation experiences of people with visual impairment, individuals who are sighted can learn how to re-understand the world through non-visual senses, explore their less utilized perceptions, empathize with the needs of the visually impaired, and recognize their artistic capabilities, thereby promoting societal equity in cultural access to art for people with visual impairment.

Accessible and equitable art education can be achieved by adopting STEAM invention. In the practice of promoting art in educational and social domains, Chao (2022) developed the "Seeing Colors" multisensory universal design educational materials. These materials created by employing STEAM knowledge translate the hue, value, and saturation of colors into text, braille, audio, visual images, and tactile textures (Figure 9.3), which makes color information accessible through touch. Circle dots represent yellow, straight lines stand for red, and wavy lines mean blue as the three primary colors. Then, the three secondary colors are mixed by two patterns of primary color. The tertiary colors are the 6 patterns with different size as the mixing proportion. This approach integrates art into various aspects of daily life, society, and culture. Thermally raised printing technology is utilized to produce dual-vision educational

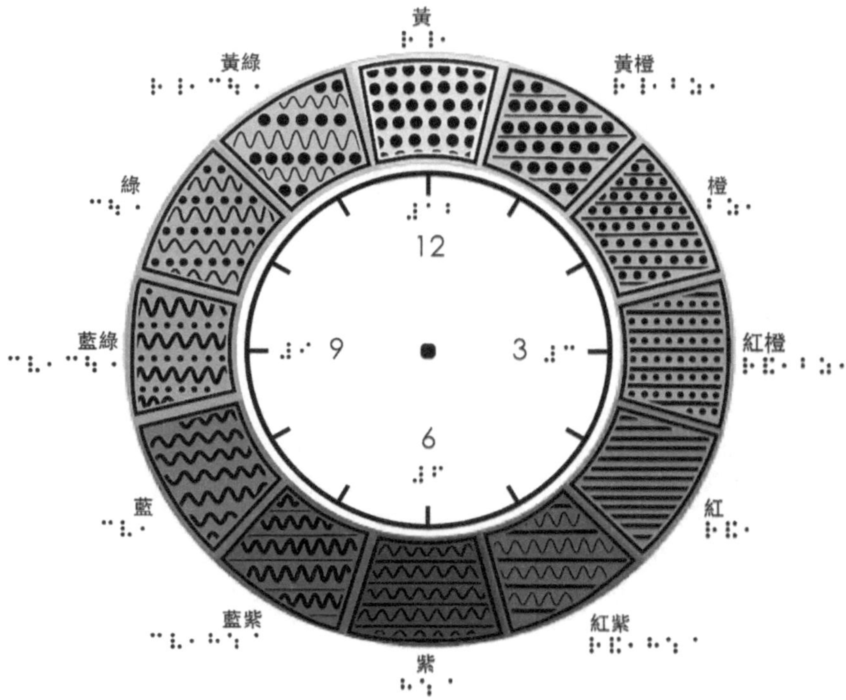

FIGURE 9.3 Universal color wheel with visual and tangible textures was invented by Chao in 2022

materials, which include units on color knowledge, art appreciation, and art creation, addressing cognitive, affective, and skills-based educational needs. These materials not only facilitate self-study for students but also find practical applications in everyday life.

As the rights of people with disabilities become more prominent and human rights issues garner increased attention, cultural equity policies are prompting public sectors to recognize these needs. The growing focus from academic research, educational curricula, civil groups, and public welfare organizations on these issues is shaping an innovative interdisciplinary field. This compels governments to consider the diverse rights of people with visual impairment, thereby extending beyond basic life skills and job training to include the right to participate in artistic and cultural activities. Educational institutions and social welfare organizations are increasingly involving people who are visually impaired in art activities, while social institutions are integrating imagery of this demographic into the media. This helps the public understand that visual impairment does not diminish individuals' rights to existence and participation.

Consequently, the societal reassessment of fine arts education for people with visual impairment, the development of specialized fine arts materials, connections with the social education system, and the creation of tactile operational aids are vital resources. These initiatives ensure the cultural and artistic rights of people who are visually impaired, thereby supporting the advancement of fine arts education tailored for their needs. This allows them to utilize diverse senses and specialized techniques for creating visual art. Through personal creative experiences, museums and galleries can offer art appreciation resources specifically designed for people who are visually impaired, thus securing their cultural rights, bolstering their creative confidence, and acknowledging their achievements. Understanding the capabilities of people with visual impairment beyond visual perception enriches human studies on non-visual sensory behaviors and performances, which adds significant potential to future cross-field research of science, technology, engineering, arts, and mathematics (STEAM) for both sighted and visually impaired in the inclusive society.

References

American Association of Museums. (1992). *The accessible museum: Model programs of accessibility for disabled and older people*. American Association of Museums.

Axel, E. S., & Levent, N. S. (2003). *Art beyond sight: A resource guide to art, creativity, and visual impairment*. American Foundation for the Blind Press.

Bandyopadhyay, S., & Rathod, B. B. (2017). The sound and feel of titrations: A smartphone aid for color-blind and visually-impaired students. *Journal of Chemical Education*, *94*(7), 946–949. https://doi.org/10.1021/acs.jchemed.7b00027

Chao, H. Y. (2016). A study on museum exhibition designs and accessible programs for visually impaired visitors. *Journal of Museum and Culture*, *12*, 105–40. https://www.airitilibrary.com/Article/Detail/P20161128001-201612-201612050009-201612050009-105-140

Chao, H. Y. (2018). Autonomy and equity: Practical application of accessible guided tour technology in the National Taiwan Museum of Fine Arts. *Journal of Museums and Culture*, *15*, 75–108. https://www.airitilibrary.com/Article/Detail/P20161128001-201806-201808280014-201808280014-75-108

Chao, H. Y. (2019). Whose photography art? Development of image interpretative methods and accessible displays for the visually impaired and blind. *Museology Quarterly*, *33*(3), 43–67. https://doi.org/10.6686/MuseQ.201907_33(3).0003

Chao, H. Y. (2020). From invisible to visible: Building the spatial cognition of museum architecture for the audiences with visual impairments. *National Taiwan Science,*

39(2), 18–29. https://www.airitilibrary.com/Article/Detail/P20150629002-202006-202011030010-202011030010-18-29

Chao, H. Y. (2022). *Seeing colors: Universal design material with multi-sensory.* Taiwan Art Beyond Vision Association.

Chen, C. L. & Chang, Y. Y. (2013). Museums and the cultural rights of the disabled: Case studies of the U.K. and Taiwan. *Museology Quarterly, 26*(2), 89–109. https://doi.org/10.6686/MuseQ.201204_26(2).0006

Chiu, T. S. (2013). Who Is the blind? Assessment, categorization, and the making up of the blind in modern Taiwan. *Taiwanese Journal for Studies of Science, Technology and Medicine, 16*, 11–47. https://doi.org/10.6464/TJSSTM.201304_(16).0001

Chiu, T. S. (2016). A Tactile turn: Exhibition ideas of Kojiro Hirose. *Journal of Museum & Culture, 12*, 141–156. https://www.airitilibrary.com/Article/Detail?DocID=P20161128001-201612-201612050009-201612050009-141-156

Hasper, E., Windhorst, R. A., Hedgpeth, T., Tuyl, L. V., Gonzales, A., Martinez, B., & Baluch, D. P. (2015). Methods for creating and evaluating 3D tactile images to teach STEM courses to the visually impaired. *Journal of College Science Teaching, 44*(6), 92–99. https://www.jstor.org/stable/43632001

Lahav, O., Hagab, N., Talis, V., & Levy, S. T. (2019). Computer-model-based audio and its influence on science learning by people who are blind. *Interactive Learning Environment*, 27(5), 856–868. https://doi.org/10.1080/10494820.2018.1500378

Ministry of Education. (2023). *Department of Statistics, the list of schools by grade and changes for school year 112.* https://depart.moe.edu.tw/ED4500/News_Content.aspx?n=63F5AB3D02A8BBAC&sms=1FF9979D10DBF9F3&s=E23C5A6CA17DB8E2

Ministry of Health and Welfare. (2023). *Number of physically and mentally challenged persons by category and county/city.* https://dep.mohw.gov.tw/dos/cp-5224-62359-113.html

Metatlav, O. (2017). Uncovering challenges and opportunities of including children with visual impairments in mainstream schools. *Proceeding of Electronic Visualisation and the Arts*, 1–6. https://doi.org/10.14236/ewic/HCI2017.102

Su, C. W., Zhuang, S. Z., Wu, H. H. and Chuang, N. T. (2018). The Research of investigating the amiable guide tour of the National Taiwan Museum of Fine Arts based on seven principles of Universal Design. *Journal of Research in Special Education and Assistive Technology, 18*, 47–57. https://tpl.ncl.edu.tw/NclService/JournalContentDetail?SysId=A18036217

The Gender Equality Committee. (2023). *Number of students with physical and mental disabilities in special education at all school levels.* https://www.gender.ey.gov.tw/gecdb/Stat_Statistics_DetailData.aspx?sn=lJfP2d02jg3xheeDsGOZ0Q%40%40

World Blind Union. (2014). *Human rights report.* https://worldblindunion.org/wp-content/uploads/2020/10/WBU-Human-Rights-Report-Nov-2014.pdf

PART 3

Reimagining STEM *as* STEAM

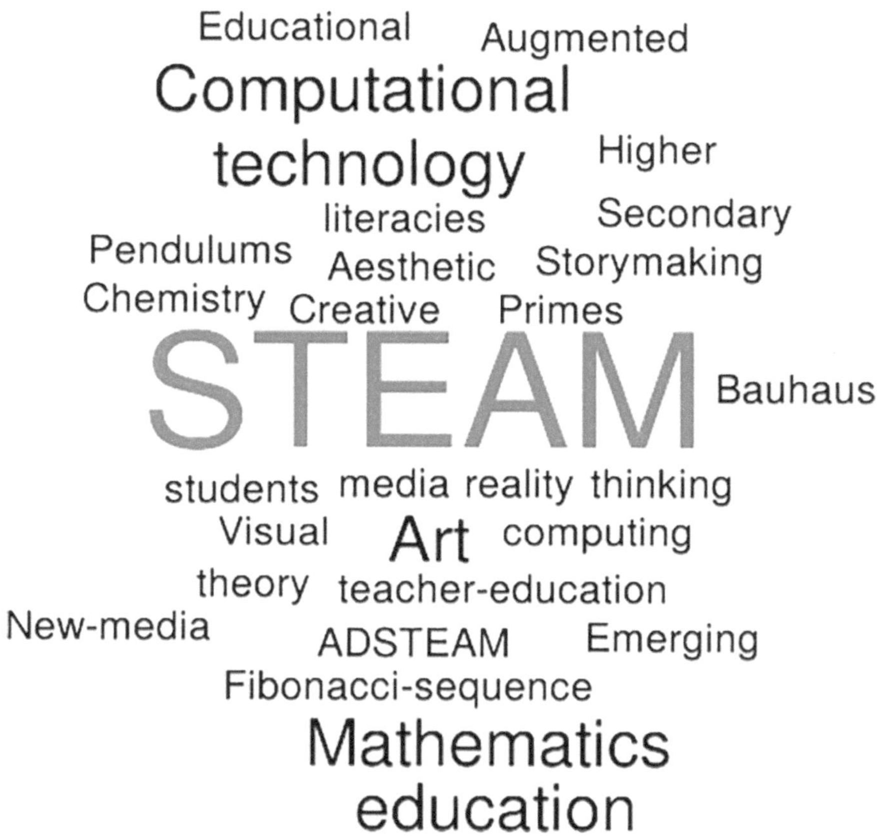
Educational
Augmented
Computational
technology
Higher
literacies
Secondary
Pendulums
Aesthetic
Storymaking
Chemistry
Creative
Primes
STEAM
Bauhaus
students
media
reality
thinking
Visual
Art
computing
theory
teacher-education
New-media
ADSTEAM
Emerging
Fibonacci-sequence
Mathematics
education

CHAPTER 10

"Uncharted Territory"

Interactive Augmented Reality for Exploring Concepts in Chemistry with Science-Averse College Students

Jonah B. Firestone and Don McMahon

Abstract

Augmented Reality (AR) allows students and teachers to overlay information from computers in the form of 3D objects, text, and animation onto reality. Learners can then interact and manipulate them in real time—individually and in groups small and large—with technology that most of them have in their pockets: smartphones. The opportunity to elucidate scientific concepts routinely misunderstood or difficult for students to grasp because of their inaccessibility has the potential to be transformative in classrooms across the content and grade spectrum. We examined the impact of this technology on science-averse preservice teachers' (n = 22) understanding of the basic concepts of chemistry. Results indicate that the incorporation of AR technology can improve understanding and engage students who normally feel disconnected from scientific content.

Keywords

augmented reality – chemistry – educational technology

1 Introduction

In recent years smartphones have become ubiquitous in daily life, yet their presence in the classroom has been at times met with disagreement. The presence of cellular phones in K–12 classrooms had been at one time viewed as a hindrance to pedagogy and a distraction to students (Baker et al., 2012; Engel & Green, 2011; Gilroy, 2004); however, with the advent of augmented reality (AR) technology, what was once considered taboo in class is now often perceived as a potential collaborative, engaging teaching tool. According to a recent survey by Common Sense Media (Kamenetz, 2019) in the United States, 53% of

 | DOI:10.1163/9789004714748_011

children by the age of 11 have smartphones as do 84% of teenagers. All Apple iPhones and most smartphones worldwide are equipped with AR capability. In other words, one of the most common objects of the 21st century can be used to display and animate content related to science, mathematics, social studies, and the arts. Often interactive, these displays and animations can bring subjects difficult to grasp or concepts hard to visualize into the classroom and onto students' desks. Because of the now commonplace nature of AR technology and the low cost of creation, educational AR experiences can be used in virtually all K–16 classrooms, incorporating technology that parents, students, and teachers already have in their pockets.

AR consists of a "direct-view" of the environment that has been enhanced by real-time computer-generated graphics or information (Carmigniani & Furht, 2011). This entails the use of some type of computer-assisted viewing device, either hand-held, such as a smartphone, or head-mounted, such as a Microsoft Hololens. In a review of the literature on AR and its use in education, Goff et al. (2018) discussed how AR differs from other types of technology like virtual reality (VR) and desktop-based 3D simulated models like Z-space. To do so, they turned to Milgram et al. (1994) and their continuum of mixed reality, which places AR on one end of the spectrum opposite VR on the other (see Figure 10.1). This distinction is important because many budding technologies may seem similar on the surface (e.g., AR and VR) but radically different in the way they are used. For instance, AR overlays the real-world environment with additional computer-generated images, but VR immerses the user in a completely computer-generated world. Although this might seem like a superficial, minor distinction, the differences between the two technologies determine how they are used for entertainment and education (Ardiny & Khanmirza, 2018).

Demonstrations, videos, models, or "hands-on" activities have been used in the sciences for over a century (Virata & Castro, 2019) to introduce students

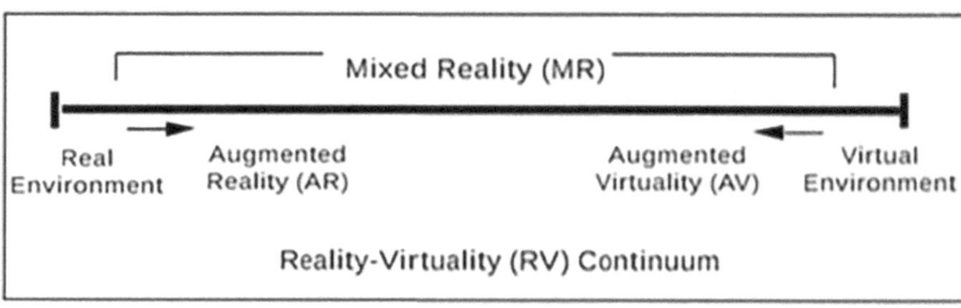

FIGURE 10.1 The reality–virtuality continuum (Milgram et al., 1994, p. 283)

to concepts difficult to visualize, such as geological time, harmonics, evolution through natural selection, and atomic bonding. These techniques have a proven success record (Flick, 1993; Johnson et al., 1997; Satterthwaite, 2010), but misunderstandings often arise because they are extremely difficult for students to fully grasp. In addition, recent studies have pointed to students who gain knowledge of skills and procedures but do not develop scientific content knowledge from these techniques (Schwichow et al., 2016). Another concern is that the challenging nature of the subject matter may lead students to a distorted view of scientific processes and concepts because of the nature of the model or activity used to introduce the concept.

Some of these issues have been addressed with AR technology. At its core AR overlays virtual objects onto the real world (Azuma et al., 2001), overlays existing either side by side with actual objects in the environment or replacing physical objects entirely. Students using AR can observe and manipulate phenomena in real time on a micro or macro scale. For example, students can "see" elements or atoms, manipulate them, and combine them as virtual 3D objects that appear on their desks or in their homes. In addition, via AR these 3D-rendered objects can be seen and shared by all members of the class or family in real time as the scientific processes under examination unfold. This sharing of experiences through AR allows students to discover phenomena and socially construct connections about science in small groups, improving understanding as hypotheses are tested in real time and can be seen and shared among students (Cheng & Tsai, 2013; Wu et al., 2013).

Not only is AR an exemplary way for students to access science content, but it can also foster creativity and art expression. Through the use of AR apps like uMake, Lightspace, and World Brush, users can draw and create digital overlays through their smartphone or tablet. These AR art applications can promote creative thinking (Muhammad, 2022) and provide users with a space to make an abstract concept more concrete. The triggers for the VR cubes that we used for this project and that are used in other cubes of this type (e.g., Mergecubes) were much more than simple QR codes; each was an intricate label with decorative designs on each side. These designs are what trigger the device to overlay the digital material. These Victorian artistic flourishes prompt the AR device to transform the static cubes into glass and brass cubes containing 3D examples of chemicals and atomic bonds.

Billingshurst et al. (2014) found that AR is overall a relatively new experience for classroom students, but since 2004 it has been shown to help students learn more effectively and increase knowledge retention relative to 2D desktop interfaces. Even simple AR scenes can be used to motivate students in a classroom because of the high level of interaction that is created, evidence Billingshurst et al. stated that AR can "enhance learning, particularly for students who learn through kinesthetic, visual, and other non-text-based methods" (p. 211).

In order to test the impact that AR might have on understanding chemical atomic bonding, along with an instructor we introduced an AR-based chemistry classroom activity into a university undergraduate basic science class and investigated the following questions:

- In what ways, if any, does the use of AR technology impact students' understanding of basic chemistry?
- How does the use of AR technology engage students in learning basic chemistry?

2 Misconceptions in Science

Understanding science and the world is based upon experience (Vygotsky, 1934/1986), which can take the form of observing first-hand, reading textbooks, watching educational or entertainment videos, and listening to classroom lectures. Previous experience can sometime include flawed understanding or errors in texts, videos, or other materials, which can lead to misconceptions (Gil-Perez & Carrascosa, 1990). An example concerns how atoms bond to create molecules, a fundamental concept in understanding the basics of chemistry. Throughout the 20th century, the evolving model of the atom (e.g., pudding model, orbit model) has produced various conceptions and misconceptions about how atoms bond to create molecules (Cokelez & Dumon, 2005; Nakhleh, 1992). Adding to the confusion is students' lack of opportunity to examine the phenomena directly (they cannot see them) or to manipulate the atoms personally (they cannot touch them) except through ball-and-stick models known to result in misconceptions; thus, students are often left with an incomplete understanding of the process or—worse—a host of misconceptions as to how atoms bond (Levy Nahum et al., 2010).

Computer simulations have been used for decades to help solve this problem (Andaloro et al., 1991; Blake & Scanlon, 2007; Ellington, 1981; Sahin, 2006). Although these simulations have been deemed somewhat helpful (Bayraktar, 2001; Rutten et al., 2012), they often replace the physical model that can be manipulated with a digital model that can be manipulated only via a mouse or other technology. These types of simulations can demonstrate factors and forces that cannot normally be seen, but they retain a degree of separation from the student, which could continue to lead to problems in understanding and create the same misconceptions that had arisen from earlier forms of instructional tools (Bayraktar, 2001).

3 AR in STEAM Education

Researchers have already begun to assess the potential of AR in the sciences. Initially, concerns about the level of programming knowledge required and time required to learn this programming prevented early adoption; however, applications like HP Reveal (formerly Aurasma; Schneider et al., 2020) have allowed teachers and researchers with limited programming skills access to the technology. For example, Plunkett (2019) detailed the ease of creating AR notecards for organic chemistry classes. These notecards, created with HP Reveal, provide interactive, animated notes to foster understanding and support retention of complicated molecules and interactions in organic chemistry. McMahon et al. (2016) used AR on mobile devices to teach medical–biology vocabulary (e.g., bones and organs and their locations in the body) to college-age students with developmental disabilities. Students in this study were able to use this intervention to interact with the AR flashcards and quickly achieve mastery of the terms and the ability to label them correctly.

Applications of AR in education are likely to continue to expand as creative platforms for AR, such as Google's ARCore and Apple's ARKit, allow anyone with a smartphone to use easy-to-access AR content and activities (Sanii, 2019); furthermore, the new Apple AR headset, the Vision Pro, brought these capabilities to users with a head-mounted display in 2024. These interactive tools have typically been precreated by companies and link to specific products or advertisements (e.g., LEGO AR-Studio). With programs, such as Blender and Halo AR, an easy-access 3D graphic design program, more complex interactive images are, however, relatively easy to create; for instance, university instructors can readily use Blender to create models of complex proteins for lessons about protein synthesis that could be uploaded to either ARCore or ARKit and then manipulated by students.

The opportunities to create AR-enabled activities are not limited to college and university science classrooms. The incorporation of this type of technology into preK–12 classrooms across content areas has already been demonstrated. Tillman et al. (2019) found that teachers participating in their study were able to create AR activities and concepts to support their existing lesson plans with minimal guidance. These lessons were not limited to science content but encompassed mathematics, literacy, art, and history. Globally, researchers have found evidence of enhanced engagement as well as improved content understanding and recall of students using AR-supported curricula (Kularbphettong et al., 2018; Rafiq & Hashim, 2018; Sanii, 2019). Overall, however, the primary focus has been on undergraduates in advanced science classes. Very little research has engaged students in entry-level science or other content classes (i.e., survey classes), nor has much research concerned students hesitant to learn science content or lacking experience in science.

4 Theoretical Frameworks: Universal Design for Learning and Constructivism

Mobile applications that use AR for any content area are prime examples of the principles of Universal Design for Learning (UDL; Walker et al., 2017). UDL is a framework of instruction in which multiple methods of engagement, interaction, and representation are emphasized (Center for Applied Special Technology [CAST], 2011). This emphasis is specifically designed to support all students in learning concepts and content that might be difficult, inaccessible, or complex. Because UDL is fundamental to the *Every Student Succeeds Act* (2015) supported by the U.S. Department of Education, it is a vital instructional concept in the preK–16 classroom. The UDL framework has three principles aligning with potential uses of AR in and outside the classroom: multiple means of (a) engagement, (b) representation, and (c) action and expression (CAST, 2011). The goal of multiple means of engagement is to create expert learners who are purposeful and motivated by providing options for recruiting interest, sustaining effort and persistence, and self-regulating. At its root, multiple means of engagement tackles the "why" of learning. The goal of multiple means of representation is to create expert learners who are resourceful and knowledgeable. This principle provides options for perception, language and symbols, and comprehension. Fundamentally, multiple means of representation support the "what" of the learning process. Finally, multiple means of action and expression provide options for physical action, expression and communication, and executive function. The goal of this principle is to create expert learners who are strategic and goal oriented. At its heart, multiple means of action and expression support the "how" of learning.

Overall, technologies such as AR in the classroom align with the framework and goals of UDL. The use of AR allows for "flexibility and equity" of use and enhances the incorporation of "perceptible information," which is at the foundation of the framework of UDL (King-Sears, 2009). Because students can control their augmented environment individually or in groups, gain content information in an unobtrusive manner, and interact with the material in a safe setting where tolerance for error is high, AR-enhanced laboratories and experiences fundamentally meet the needs that UDL was designed to address (Walker et al., 2017).

AR activities, designed to align with the UDL guidelines, can greatly enhance art education as part of science, technology, engineering, arts, and mathematics (STEAM) initiatives. By incorporating AR into art lessons, educators can provide students with immersive and inclusive learning experiences that cater to diverse learning styles and abilities.

With AR activities the UDL principle of providing multiple means of representation can be addressed. Students can use AR to explore artworks from various perspectives, examine intricate details, and witness the creative process behind them. They can visit a virtual art gallery or museum, where they can view and interact with renowned artworks and artifacts. AR activities aligned with the UDL principle of offering multiple means of expression allow students to create their own digital artwork, incorporating elements of both the physical and virtual realms. For instance, they can overlay digital images, animations, and sounds onto their traditional drawings or sculptures, bringing their creations to life in new ways. This empowers students to express their artistic ideas and concepts uniquely and personally, fostering creativity, innovation, and self-expression. AR-based art activities provide opportunities for the third UDL principle of multiple means of engagement by supporting collaborative learning: Students can work together to create shared AR experiences or engage in virtual exhibitions in which they showcase their artwork for a wider audience, transcending the traditional boundaries of the classroom.

The integration of augmented reality activities based on the UDL guidelines can support art education within STEAM initiatives. By leveraging AR technology, educators can offer students immersive and inclusive art experiences that cater to their diverse needs, preferences, and abilities. Through the multiple means of representation and expression provided by AR, students can deepen their understanding of art, refine their skills, and unleash their creativity while fostering collaboration and connection within the artistic community.

In framing our study, we took a constructionist perspective in which individuals and groups are presumed to interact together in their environment, generating meaning (Vygotsky, 1934/1986). Constructivism is "the view that all knowledge, and therefore all meaningful reality, is contingent upon human practices, being constructed in and of interaction between human beings and their world and developed and transmitted within an essentially social context" (Crotty & Crotty, 1998, p. 42). Thus, we viewed participants as interacting to find meaning and relevance and designed the classroom activity to maximize the interaction between individuals, groups, and resources (King-Sears, 2009; Vygotsky, 1934/1986). This framework is apt for studying the role of AR technology in teaching and learning because AR technology fosters direct interaction with concepts and content normally inaccessible or difficult to grasp. In addition, because of the open nature of AR images, this interaction is often done with others in small groups in which all participants can observe, manipulate, and engage with the material simultaneously to construct their own understandings. Taken as a whole, the use of AR technology aligns with both the fundamental goals of UDL and theories of experiential learning and social construction (King-Sears, 2009; Vygotsky, 1934/1986).

5 Background of the Study

In 2020, we conducted a research study to determine impact that AR might have on preservice elementary school teachers' basic understanding of the elements of two types of chemical bonding. To create a robust and full understanding of the students' responses and reactions, we performed an initial quantitative analysis of change from preassessment to postassessment and conducted a subsequent qualitative analysis of interviews with the instructor of the class.

5.1 *Setting*

The AR technology was introduced during a general science survey course at a regional campus of a large research university in the Pacific Northwest of the United States. We designed the course to introduce students to various subjects in earth science, physics, chemistry, and environmental science. The College of Arts and Sciences for the university houses the course as part of its overall curriculum; however, the curriculum and pedagogical standards are overseen by the College of Education at the university as part of the certification program for preservice K–8 teachers at the college. The instructor was an adjunct professor who taught the lecture and laboratory sections under the guidance of Dr. Firestone, a science education professor in the College of Education.

We used the Elements 4D AR application developed in 2014 by DAQRI, a California-based AR company, now defunct. Information about and a demonstration of this application can be viewed elsewhere online ("Elements 4D Interactive Blocks," 2024). The application projected a 3D image of an element, which viewed through an AR-enabled device, such as an iPhone or iPad, appeared to be contained in a glass box. The projected image of the box was overlaid on a paper cube available for printing on the developer's website. Each cube had the name of an element, the corresponding chemical symbol, and the atomic weight on each of its six faces. In addition, on each face of the cube was an artistic design that the AR program used to identify the element. There were six cubes in each set, so in total, each set contained 36 elements.

Whichever element was on the top of the cube at the time determined the element that was displayed via the AR application. If a student turned the cube and a different element appeared on the top of the paper cube, the projected element changed.

The projected images of the elements inside glass boxes were in whatever phase (solid, liquid, or gas) in which they appear at room temperature. If someone moved around the cube and looked through an AR-enabled device, the 3D glass cube remained stationary while the viewer's point of view changed. Because of this, when viewed through a device, the elements in the glass boxes appeared to exist in the environment. At the same time, a viewer could reach out and move the cube, turn it at various angles, or flip it over; and the elements and the glass box moved, turned, or flipped over.

Within the AR application, viewers could select additional information about the element displayed, including a brief description, melting and freezing points, and its location on the periodic table. When another cube from the same set of six was introduced, the projected image also appeared as whatever element was its top face. If the two cubes touched and the elements were able to bond, a new image of a glass box containing the new substance appeared, replacing the two original boxes. For instance, when a viewer placed the box with solid sodium next to the box with chlorine gas, both disappeared and were replaced with a glass box containing salt (NaCl). In addition, the viewer could select the new substance and receive information about its qualities. If two boxes placed next to each other with elements were unable to bond, then both remained unchanged.

5.2 *Participants*

The participants (n = 22), including 17 women and 5 men, were evenly split between White and Latinx students. All were undergraduates at the university and 19 were enrolled as preservice teachers in the College of Education; three had not yet declared their major. None of the students involved had prior college experience in chemistry or other college-level science classes. All the students verbally reported some knowledge of or experience with AR technology either through playing games like Pokémon Go in the past or having friends who had shared a similar experience with AR.

5.3 *Procedures*

The research design was a pre- and posttest one-group observational design. Although no comparison group was present, the principal advantage of this design is the possibility it presents to examine changes in the dependent variable. When used in conjunction with an effect-size measure, identifying portions of the causal relationship between the intervention and the change in scores is possible. We created the pre- and postassessment with input from the instructor. Both assessments comprised a combination of 20 multiple choice and short answer questions drawn from an introductory high school chemistry text. The assessments included three questions on basic atomic vocabulary, five questions about the periodic table, three questions that involved balancing chemical equations, four questions about atomic bonding, and five questions about types of chemical reactions. At the beginning of the 3-hour activity, the instructor administered the students the preassessment. It had no time-limit for completion; however, all the students finished within 25 minutes.

After the preassessment, the students moved into their six previously established activity groups of four students each (two students were missing, leaving two groups with only three members apiece). Six Elements 4D cubes were placed at each lab table. The cubes were plain wooden blocks with the Elements 4D symbols taped to each side. The instructor provided an iPad mini with the Elements 4D cube program preloaded to each of the groups.

The instructor used a modified version of a Grades 6–8 curriculum provided with the AR application. We conferred with the instructor and chose the middle school curriculum because of its introductory nature. The students had little previous experience in chemistry, and many demonstrated trepidation about engaging in science. In addition, because the science class had been designed with preservice elementary teachers in mind, we and the instructor felt that the curriculum provided a low-stakes introduction to elements and chemical bonding. In the future, we thought, the preservice teachers could use the format as a basis for their own science curriculum.

The class participated in a brief, instructor-led discussion about the properties and states of matter and how chemists use color, state, and volume to classify and organize elements. The activity groups then spent the next 20 minutes engaging with the AR applications and the cubes, discovering how the technology worked, asking questions about AR and the projected elements, and reading the information provided via the application about each element. At the same time, we and the instructor moved around, answering questions about the technology and fixing any problems with cubes or the application. In their activity groups, the students filled out a worksheet, listing each of the 36 elements in their set, its location on the periodic table, color, state of matter at room temperature, and any additional information they deemed interesting about the element. Afterward, the instructor led a short debrief, then the students took a break.

When they returned, the instructor led a brief whole-class discussion about what they learned from the cubes about the periodic table. They discussed any patterns they noted from the previous exercise about how elements were grouped and any specific qualities of elements within these sets. The students then broke into their activity groups and worked on discovering which elements could bond and which could not (within each set 10 possible compounds could be created). The students filled out an additional worksheet listing the name of each element, group with which it could bond, and the compound that was then formed. When all the student groups had found and identified the 10 possible compounds, the instructor called the students together and delivered a lecture on various types of chemical reactions, the role of valence electrons in bonding, ionic and covalent chemical bonds, and different types of chemical reactions. In their activity groups the students discussed and recorded what they believed to be the chemical reactions that occurred to create the 10 compounds and then shared their ideas with the rest of the class as part of a whole-class discussion. The instructor then debriefed the class and answered any questions about chemical bonding and balancing simple chemical equations.

5.4 *Results*

At the beginning of the activity session two weeks later, the instructor administered the postassessment to the class. We then scored the pre- and postassessments. After the activities and assessments, we interviewed the instructor about his observations and opinions about the technology, the activities, and the results.

5.4.1 Quantitative

In order to evaluate any impact of the intervention on student understanding, we conducted a dependent *t*-test to evaluate whether a statistically significant positive difference existed between pre- and postassessment outcomes. The test was significant, $t(21) = 3.13$, $p = 0.03$, $d = .77$, suggesting a statistically significant positive difference, indicating increased correct answers (or better learning outcomes) in posttest. We also conducted a Mann–Whitney U test to determine whether any statistical differences were based upon gender or ethnicity in the class. The distribution of the change score across the category of gender was the same ($U = 19.00$, $z = -1.85$, $p = .065$), revealing no difference between men and women. The distribution of the change score across the category of ethnicity was also the same ($U = 52.00$, $z = -.560$, $p = .575$), pointing to a lack of difference between ethnicities. Overall findings suggest that students' posttests were better than pretests regardless of their gender and ethnicity.

In terms of individual questions, change was uneven across content type. The five questions about the periodic table and how it related to chemical bonds showed both the greatest change in the ratio of wrong to right answers per question type (four of five questions) and the greatest gain on any specific question (a gain of eight right answers). All three of the questions about chemical equations showed some improvement across the class. Of the three questions concerning vocabulary, two increased and one decreased from pre- to posttest. The four questions about types of bonding were the most mixed, with two remaining the same, one improving, and one declining in the number of students with the correct answer from pre- to posttest. On the four questions about types of chemical bonding and the five questions about types of chemical reactions, students did the poorest from pretest to posttest, improving only on two of five questions about reactions, staying the same on one question, and declining on two (see Table 10.1).

TABLE 10.1 Overall change per question and question type from pretest to posttest

Question	Type	Change: pre–post
1	Vocabulary	2
2	Vocabulary	−1
3	Vocabulary	3
4	Table	5
5	Table	4
6	Table	−2
7	Table	8
8	Equations	2
9	Equations	1
10	Bonding	0
11	Bonding	0
12	Bonding	4
13	Reactions	−2
14	Reactions	1
15	Reactions	0
16	Reactions	−4
17	Reactions	2
18	Equations	5.34
19	Table	5
20	Bonding	−1

5.4.2 Qualitative

The instructor shared several valuable insights into the ability, growth, and engagement of the students during the AR activities. During the interview with the instructor (personal communication, Feb. 8, 2019), he indicated three key ways that the use of the Elements 4D Cubes seemed to enhance the students' learning experience. First, the use of the cubes provided a bridge to unfamiliar content. Second, the cubes lent themselves to a visual, interactive mode of learning. Finally, he noted that the students were much more engaged than he normally observed them to be and seemed to enjoy working with the AR. The students were able to "see" the elements and "observe" the bonding of the elements to create compounds in a way that was radically different from their previous experiences in science in general and in chemistry in particular.

Bridging the Gap. One of the key observations shared by the instructor was the way in which manipulating the cubes helped to bridge the gap in knowledge with which many students entered class. Describing his initial expectation that students lacked the background knowledge needed for the lab, he said:

> I remember thinking that there was going to be a measurable distance in their knowledge base .... Most of our students do not have a strong science background. They did not have an idea of what was going on. It was uncharted territory.

Because of this gap in knowledge, he also noted how students tended to hold negative feelings toward science and often felt overwhelmed by the content of the course: "Every semester I ask the students what their background in science is because I like to get an idea, and overwhelmingly it is negative." During the activity with the AR cubes, however, the information needed to successfully complete the tasks in the lesson was embedded within the manipulatives themselves. In the first phase of the lesson, the students were encouraged to explore the 36 elements. They could do so in any sequence they wanted—putting them in charge of their learning. The instructor noted that this allowed for a more intuitive experience. Students were able to access the necessary information at the moment they needed it. Holding the AR scanner above the cube generated a read-out of information about each element. This innate function of the cubes—to provide the information as needed—allowed novices to be successful.

Active Learning Leads to Greater Understanding. Another aspect of the engagement was related to the ability to manipulate the cubes, which according to the instructor facilitated learning. During the second phase of the lesson, students were able to test out various combinations of elements to see which ones bonded and which ones did not. This ability to test out their hypotheses and receive real-time, in-the-moment feedback allowed students to monitor and adjust their developing understandings. He stated that this led to greater understanding, noting, "They learn better by having some sort of tactile … exercise. They would have a higher level of understanding rather than just reading about the activity or watching a video."

The instructor also related this interactive style of learning to the students' future vocation, likening it to a K–12 setting, and using playing with clay as a metaphor for this hands-on approach. He reflected, "They are education majors. They learn by doing; they learn by manipulating clay." He thought that this led to deeper understanding and lamented the inability to compare their learning to that of a group learning the same content without the use of the 4D cubes, noting, "It would have been interesting to have a control group. I believe that the people using the tool would have a higher understanding than those who did not." We did not have the opportunity to test this out; however, the instructor's belief that this AR-enhanced laboratory facilitated a high level of understanding in his students was noteworthy.

Engagement. Another key observation related to the way the AR technology fostered student engagement. The instructor attributed this engagement to the tactile and visual nature of the cubes and the way they required manipulation to see the various elements and how they bonded or did not bond. He recalled:

> I remember getting into the lab and thinking how the simulated tactile learning, the combination and transformation of paper blocks into visual animation that they could manipulate really turned people on. They were giggling, they were asking questions, they seemed really engaged with the tool.

Not only did students engage with the materials, but they also seemed to genuinely enjoy the investigation, demonstrated by their laughter. They also asked questions, another sign of engagement with the material. Overall, the use of AR technology appeared to have several benefits, perhaps not the least of which was enjoyment of the material. The instructor noted, "So, when you can get students who are largely averse to science to have an activity relating to chemistry and say that they enjoyed it, it is a monumental step."

6 Discussion, Limitations, and Conclusions

The results of this study reveal several benefits and constraints of using AR in the classroom. Below is a discussion of the results of the two research questions.

First, in what ways, if any, does the use of AR technology impact students' understanding of basic chemistry? Overall, the students improved between pre- and posttest, not necessarily surprising because the activity was directly related to the content of the assessment; however, the areas in which the students did and did not improve are more telling. The students showed the greatest improvement on questions directly related to the 4D cubes: location on the periodic table and how position related to chemical bonds as well as equations and bonding. Each of these concepts was demonstrated by the activities with the 4D cubes. Some concepts were demonstrated on one hand by passively showing the location and identification of the type of element (metal, nonmetal, metalloid) on the application (information that appeared around the element on the application) and on the other hand by actively depicting which elements bonded to create a new compound and which did not (appearing when the students correctly placed the compound forming elements together).

This contrasts with the concepts with which the students improved the least from pre- to posttest: types of bonding and chemical reaction types. These subjects were not covered directly on the 4D application. The instructor discussed them in a whole-group setting between activities with the cubes, and the students took notes and asked questions about these subjects; however, except for some drawings on the white board, the students did not have graphic or physical examples or models, nor did they engage in active manipulation as part of the learning of these concepts. In addition, the questions concerning vocabulary demonstrated only some improvement (two of three). Two of the questions asked about "the smallest unit of matter" and "elements which are malleable and can be pressed into sheets [and] are mostly solid at room temperature." Both questions were either directly discussed in the 4D application text or shown as part of the displayed image (the metals were shown as solid in the glass cubes). The vocabulary question on which the students did not improve—"Two or more chemically bonded atoms are a(n) _____"—was not referenced in the 4D application nor the lecture portions of the activities.

Second, how does the use of AR technology engage students in learning basic chemistry? The instructor noted several aspects of student behavior that directly related to the way they had benefited from using the cubes. The students bridged the gap between information presented in the cubes and knowledge that they could use on the assessment. They initially had very tentative knowledge of chemistry, chemical bonding, and the periodic table; however, according to the instructor, after spending some time interacting with the cubes, they seemed much more confident in the content. In addition, because the content was embedded in the cubes themselves, perceived barriers to learning subjects that they found intimidating in the past seemed lower.

In terms of UDL, the 4D cubes provided a low-stakes avenue to learn about chemistry. The AR activity provided the students with "perceptible information" through tactile manipulation of the cubes. At the same time, the application provided "supporting details and precise and clear language" (King-Sears, 2009, p. 200) that fostered learning and engagement for the students, who were able to flexibly use the cubes individually and in groups, keeping them engaged throughout the experience. Throughout the AR classroom activity, the students were able to use multiple means of engagement, representation, action, and expression to gain understanding. Despite the challenges the students experienced in previous labs, in terms of aversion and lack of motivation, as reported by the instructor, the use of the cubes fostered understanding and engagement in alignment with the goals of UDL (Walker et al., 2017).

At the same time, the activities were learning labs unto themselves. The students were able to actively test out theories about what could and could not bond in real time, which they are unable to directly observe in more mainstream chemistry activities. With the cubes, the rules were embedded for the students to discover themselves. Although they were likely aware of some combinations, such as how hydrogen bonds with oxygen to create water, because of the "augmented" nature of the activities, the students could "experience" this type of bonding firsthand in a way that textbooks, stick-and-ball models, and other types of chemistry teaching cannot demonstrate.

Overall, the subjects in which the students engaged physically, observed, and read via the AR application were better learned or retained during the two-week interval between the pretest and the posttest. At the same time, students were more engaged with the content, less averse to the subject matter, and showed interest in the prospect of conducting AR-enabled classroom activities. Ultimately, student manipulation of and interaction with various scientific concepts, such as atomic bonding, using AR technology, can lead to a greater understanding of this and other challenging science concepts. Students who had been averse to science became engaged in the subject matter in a low-stakes, low-barrier environment that included paper cubes with artistic flourishes. The use of AR technology was motivating and interesting to them; thus, the students who conducted the AR laboratory demonstrated an interest in science that they had not shown before nor reported having.

Neither the technology nor the content alone can make this type of impact by itself, but when combined, the students can create knowledge and meaning in ways that were impossible just a few years ago. AR technology and other associated technologies such as VR are not meant to replace teacher-led instruction, traditional hands-on laboratories, or demonstrations; however, they can act to "augment" current methods of instruction in science and other content areas to engage students in different ways, buttress existing knowledge, help reduce misconceptions, and support students who are averse to science or other subjects because of past experiences.

References

Andaloro, G., Donzelli, V., & Sperandeo-Mineo, R. M. (1991). Modelling in physics teaching: The role of computer simulation. *International Journal of Science Education, 13*(3), 243–254. https://doi.org/10.1080/0950069910130303

Ardiny, H., & Khanmirza, E. (2018, October). The role of AR and VR technologies in education developments: Opportunities and challenges. In *2018 6th RSI International Conference on Robotics and Mechatronics* (pp. 482–487). Institute of Electrical and Electronics Engineers.

Baker, W. M., Lusk, E. J., & Neuhauser, K. L. (2012). On the use of cell phones and other electronic devices in the classroom: Evidence from a survey of faculty and students. *Journal of Education for Business, 87*(5), 275–289. https://psycnet.apa.org/doi/10.1080/08832323.2011.622814

Bayraktar, S. (2001). A meta-analysis of the effectiveness of computer-assisted instruction in science education. *Journal of Research on Technology in Education, 34*(2), 173–188. https://doi.org/10.1080/15391523.2001.10782344

Billinghurst, M., Clark, A., & Lee, G. (2014) A survey of augmented reality. *Foundations and Trends in Human–Computer Interaction, 8*(2–3), 73–272. http://dx.doi.org/10.1561/1100000049

Blake, C., & Scanlon, E. (2007). Reconsidering simulations in science education at a distance: Features of effective use. *Journal of Computer Assisted Learning, 23*(6), 491–502. https://doi.org/10.1111/j.1365-2729.2007.00239.x

Carmigniani, J., & Furht, B. (2011). Augmented reality: An overview. In B. Furht (Ed.), *Handbook of augmented reality* (pp. 3–46). Springer. https://doi.org/10.1007/978-1-4614-0064-6_1

Center for Applied Special Technology. (2011). *Universal design for learning guidelines. Version 2.0*. Author.

Cheng, K. H., & Tsai, C. C. (2013). Affordances of augmented reality in science learning: Suggestions for future research. *Journal of Science Education and Technology, 22*(4), 449–462. http://dx.doi.org/10.1007/s10956-012-9405-9

Cokelez, A., & Dumon, A. (2005). Atom and molecule: Upper secondary school French students' representations in long-term memory. *Chemistry Education Research and Practice, 6*(3), 119–135.

Crotty, M., & Crotty, M. F. (1998). *The foundations of social research: Meaning and perspective in the research process*. Sage.

Elements 4D interactive blocks. (2024). *Kickstarter*. https://www.kickstarter.com/projects/daqri/elements-4d-interactive-blocks

Ellington, H. (1981). *Games and simulations in science education*. Nichols Publishing Company.

Engel, G., & Green, T. (2011). Cell phones in the classroom: Are we dialing up disaster? *TechTrends*, *55*(2), 39–45. https://doi.org/10.1007/s11528-011-0482-z

Every Student Succeeds Act, 20 U.S.C. § 6301 (2015). https://www.congress.gov/bill/114th-congress/senate-bill/1177

Flick, L. B. (1993). The meanings of hands-on science. *Journal of Science Teacher Education*, *4*(1), 1–8. http://dx.doi.org/10.1007/BF02628851

Gil-Perez, D., & Carrascosa, J. (1990). What to do about science "misconceptions." *Science Education*, *74*(5), 531–540. https://psycnet.apa.org/doi/10.1002/sce.3730740504

Gilroy, M. (2004). Invasion of the classroom cell phones. *The Education Digest*, *69*(6), 56–60.

Goff, E. E., Mulvey, K. L., Irvin, M. J., & Hartstone-Rose, A. (2018). Applications of augmented reality in informal science learning sites: A review. *Journal of Science Education and Technology*, *27*(5), 433–447. https://www.learntechlib.org/p/189101/

Johnson, D. M., Wardlow, G. W., & Franklin, T. D. (1997). Hands-on activities versus worksheets in reinforcing physical science principles: Effects on student achievement and attitude. *Journal of Agricultural Education*, *38*(3), 9–17. https://doi.org/10.5032/jae.1997.03009

Kamenetz, A. (2019, October 31). It's a smartphone life: More than half of US children now have one. *National Public Radio: Ideastream Public Media* https://www.npr.org/2019/10/31/774838891/its-a-smartphone-life-more-than-half-of-u-s-children-now-have-one

King-Sears, M. (2009). Universal design for learning: Technology and pedagogy. *Learning Disability Quarterly*, *32*(4), 199–201. https://doi.org/10.2307/27740372

Kularbphettong, K., Roonrakwit, P., & Chutrtong, J. (2018, July). Effectiveness of enhancing classroom by using augmented reality technology. In *International Conference on Applied Human Factors and Ergonomics* (pp. 125–133). Springer. http://dx.doi.org/10.1007/978-3-319-93882-0_13

Levy Nahum, T., Mamlok-Naaman, R., Hofstein, A., & Taber, K. S. (2010). Teaching and learning the concept of chemical bonding. *Studies in Science Education*, *46*(2), 179–207. http://dx.doi.org/10.1080/03057267.2010.504548

McMahon, D. D., Cihak, D. F., Wright, R. E., & Bell, S. M. (2016). Augmented reality for teaching science vocabulary to postsecondary education students with intellectual disabilities and autism. *Journal of Research on Technology in Education*, *48*(1), 38–56. http://dx.doi.org/10.1080/15391523.2015.1103149

Milgram, P., Takemura, H., Utsumi, A., & Kishino, F. (1994). Augmented reality: A class of displays on the reality–virtuality continuum. *Telemanipulator and Telepresence Technologies*, *2351*, 282–292. http://dx.doi.org/10.1117/12.197321

Muhammad, M. (2022, November). Review of trends in learning media of augmented reality integrated with STEM approach to improve students' creative thinking skill. *Journal of Physics: Conference Series, 2377*(1), http://dx.doi.org/10.1088/1742-6596/2377/1/012084

Nakhleh, M. B. (1992). Why some students don't learn chemistry: Chemical misconceptions. *Journal of Chemical Education, 69*(3), 191–196. http://dx.doi.org/10.1021/ed069p191

Plunkett, K. N. (2019). A simple and practical method for incorporating augmented reality into the classroom and laboratory. *Journal of Chemical Education, 96*(11), 2628–2631. https://doi.org/10.1021/acs.jchemed.9b00607

Rafiq, K. R. M., & Hashim, H. (2018). Augmented reality game (ARG), 21st century skills and ESL classroom. *Journal of Educational and Learning Studies, 1*(1), 29–34. http://dx.doi.org/10.32698/0232

Rutten, N., Van Joolingen, W. R., & Van Der Veen, J. T. (2012). The learning effects of computer simulations in science education. *Computers & Education, 58*(1), 136–153. https://psycnet.apa.org/doi/10.1016/j.compedu.2011.07.017

Sahin, S. (2006). Computer simulations in science education: Implications for Distance Education. *Turkish Online Journal of Distance Education, 7*(4), Article 12. https://files.eric.ed.gov/fulltext/ED494379.pdf

Sanii, B. (2019). Creating augmented reality USDZ files to visualize 3D objects on student phones in the classroom. *Journal of Chemical Education, 97*(1), 253–257. https://doi.org/10.1021/acs.jchemed.9b00577

Satterthwaite, D. (2010). Why are 'hands-on' science activities so effective for student learning? *Teaching Science, 56*(2), 7–10 (EJ907322). ERIC. https://eric.ed.gov/?redir=http%3a%2f%2fwww.asta.edu.au%2fresources%2fteachingscience Schneider, J., Patfield, M., Croft, H., Salem, S., & Munro, I. (2020). Introducing augmented reality technology to enhance learning in pharmacy education: A pilot study. *Pharmacy, 8*(3), 109. https://doi.org/10.3390%2Fpharmacy8030109

Schwichow, M., Zimmerman, C., Croker, S., & Härtig, H. (2016). What students learn from hands-on activities. *Journal of Research in Science Teaching, 53*(7), 980–1002. https://doi.org/10.1002/tea.21320

Tillman, D., Alvidrez-Aguirre, V., Kim, S. J., & An, S. (2019). Teachers' conceptions of the pedagogical potential for classroom-based augmented reality. *Journal of Educational Multimedia and Hypermedia, 28*(4), 411–434. https://www.learntechlib.org/primary/p/182417/

Virata, R. O., & Castro, J. D. L. (2019, January). Augmented reality in science classroom: Perceived effects in education, visualization and information processing. In *Proceedings of the 10th International Conference on E-Education, E-Business, E-Management and E-Learning* (pp. 85–92). https://doi.org/10.1145/3306500.3306556

Vygotsky L. (1986). *Thought and language.* (A. Kozulin, Trans.). MIT Press. (Original work published 1934)

Walker, Z., McMahon, D. D., Rosenblatt, K., & Arner, T. (2017). Beyond Pokémon: Augmented reality is a universal design for learning tool. *SAGE Open*, *7*(4). https://doi.org/10.1177/2158244017737815

Wu, H.-K., Lee, S. W.-Y., Chang, H.-Y., & Liang, J.-C. (2013). Current status, opportunities and challenges of augmented reality in education. *Computers & Education*, *62*, 41–49. https://doi.org/10.1016/j.compedu.2012.10.024

CHAPTER 11

Incorporating Art in Mathematics Teacher Education

Examples with Primes and Pendulums

David Glassmeyer and Kevin Hsieh

Abstract

This chapter details two ways we have incorporated art into the mathematics education of in-service mathematics teachers. The first example allows teachers to visualize whole numbers as configurations of shapes that highlight features of prime versus composite numbers. The second example allows teachers to explore the relationships of various features of a swinging pendulum, reason about the mathematical relationships between observed quantities, and explain visually pleasing wave motions using these relationships.

Keywords

mathematics – mathematics teacher education – primes – pendulums – secondary students – visual art

1 Introduction

A shortage of graduates in science, technology, engineering, and mathematics (STEM) subjects worldwide has prompted integrated educational approaches to better support student interests, understanding, and motivation to pursue them (National Science Board, 2022; Tayebi et al., 2021). More recently, the arts have been included, integrating science, technology, engineering, the arts, and mathematics (STEAM) education to improve the support of student creative thinking and interdisciplinary skills needed in the modern workforce (Aguilera & Ortiz-Revilla, 2021; Suganda et al., 2021). More specifically, the National Art Education Association (NAEA) has defined the STEAM approach as "the integration of the arts and design principles, concepts, and techniques with STEM

 | DOI:10.1163/9789004714748_012

instruction and learning" (2022, para. 1). An example of secondary STEAM education can be found in our earlier work, in which students used engineering, mathematics, and cultural concepts to construct lanterns in visual art classes (Glassmeyer et al., 2018; Hsieh et al., 2019).

Visual art can play a supportive role in motivating students to learn and understand STEM content knowledge, with teachers playing a key role in implementing STEAM education by significantly affecting student interest in and understanding of content knowledge (Autenrieth et al., 2017; Brophy et al., 2008; Nutov, 2021). Integrated STEAM education is challenging for teachers, requiring cross-disciplinary collaboration, increased planning and work, and an understanding of STEAM content connections (Boice et al., 2021). One established method of supporting teachers to successfully integrate STEAM into their classrooms is through professional development (PD) efforts. "STEAM PD may be most successful when it involves collaborative planning time for teachers across disciplines, a shared pedagogical foundation (such as PBIL [problem-based inquiry learning]), and hands-on opportunities for teachers to experience STEAM learning" (Boice et al., 2021, p. 18); however, more work is needed to detail how to support teachers in developing an understanding of integrated STEAM content.

In this chapter, we detail two lessons used in STEAM PD to help develop mathematics teachers' understanding of how the arts can be incorporated in secondary mathematics classrooms. The first lesson, called Primes, involves visualizing whole numbers as configurations of shapes that highlight features of prime versus composite numbers, providing an easy introduction to mathematical topics of factors and multiples. In the second lesson, called Pendulums, a swinging pendulum is used to prompt teachers to experiment with various features (e.g., rod length, string length, weight) to identify quantities and establish mathematical relationships between them. We begin by describing features of STEAM PD guiding the design and implementation of the lessons.

2 Frameworks for Integrated STEAM Learning

Frameworks by Henriksen et al. (2019) and Perignat and Katz-Buonincontro (2019) guided our process of integrating STEAM learning for mathematics teachers. They recommended using curiosity to explore the interplay between mathematics and non-mathematics ideas and then making those connections accessible for teachers and their students, usually involving technology or visual aids that facilitate student exploration or simplification of STEM concepts.

Incorporating these recommendations when designing and implementing the two lessons described in this chapter, we used technology to provide students with interactive websites, allowing them to explore patterns within prime numbers and experiment with the impact of several attributes on the swinging of a virtual pendulum. Visual aids including diagrams of pendulum swinging patterns were also helpful to students in visualizing the patterns. The exploration and experimentation fostered student curiosity about the patterns they saw, prompting a need to use mathematics concepts to explain the patterns observed. Students were thus motivated to understand and explain the visual arts and to explore the interplay of mathematics concepts at work. These lessons were implemented with teachers in a STEAM PD setting but could be used in a graduate-level course with in-service teachers or in undergraduate accreditation programs with preservice teachers. With some adaptation, these activities could also be used by high school teachers with their secondary students.

3 STEAM Lesson 1: Primes

Prime numbers are included in the curriculum for the elementary grades, and several topics for middle and high school rely on factors and multiples as well; therefore, this two-part Primes STEAM Lesson can serve as a (re)introduction to these topics for middle and secondary mathematics students. In Part 1, learners observe an animation of the numbers. In Part 2, their attention is directed toward the placement of the primes in relation to multiples of six, highlighting how every prime larger than 3 is either one larger or one smaller than a multiple of six, a pattern often undiscovered in secondary mathematics (Neale, 2017).

3.1 *Part 1: Exploring How Numbers Can Be Represented*

Part 1 begins with students viewing the Animated-Factorization-Diagrams website (Von Worley, 2012), which provides a visualization of the natural numbers arranged by their prime factorization that automatically progresses from one to the next. For example, the number 77 on the website shows 11 circles comprising seven dots in each circle, relating to $77 = 11 \times 7$. This website was developed by Von Worley (2012) based on a variation of Yorgey's (2012) factorization diagrams, but to our knowledge, it has not been previously incorporated into educational lessons.

We invited teachers to let the animation run or click the fast forward button, then the pause button at the bottom right of the Animated-Factorization-Diagrams website to find a number whose visualization they thought was interesting. Then we asked teachers to answer the following questions.

- What patterns do you notice?
- How does the visual representation relate to the factorization of the number?
- Can you determine the factorization based only on the visual representation?
- Could you draw the visual representation based only on the factorization?
- What rules does this website use in terms of grouping the dots when representing a number? More specifically, how do you think the person who made the website programmed the dots to be arranged?
- Find a number that you find visually pleasing. What aspects of the visualization do you like, and how do they relate to the factors of that number?
- You have probably noticed that when the word "prime" appears, that number is represented as a single arrangement of dots in a circle. How often do these prime numbers (or single arrangements of dots in a circle) appear?

Teachers found that they could draw the visualization based on the prime decomposition of the number and that prime numbers seemed to appear less frequently as the numbers grew larger.

3.2 *Part 2: Exploring Patterns in Number Representations*

After Part 1 we provided an image of the factorization diagrams for the first 36 natural numbers in a single visualization (Yorgey, 2012), using six columns and six rows. Then we asked teachers to discuss in small groups the following questions:

- After the first row in which columns do the prime numbers appear? Remember, a prime number is represented by a single arrangement of dots in a circle.
- After the first row which columns will never have a prime number? How can you justify or prove this?

Teachers discussed reasons that not every number in Column 1 was prime and reasons that not every number in Column 5 was prime but all the primes greater than 3 occurred in these columns. They justified this by describing why every number in Columns 2, 4, and 6 are multiples of 2, and every number in Column 3 is a multiple of 3. Note that every number in Column 5 is not a multiple of 5 (e.g., the number 10 falls in Column 4, the number 15 falls in Column 3, etc.).

3.3 *Primes Lesson Discussion*

This lesson, which incorporated visual art to depict prime numbers, invited teachers to use these visual representations of factors to notice and justify patterns in the primes. Specifically, a single circle of dots represented the prime numbers because these numbers could not be represented as the product of factors, which were the grouping representation of the nonprime numbers (called composite numbers). We found the teachers enjoyed this part of the lesson, with several teachers sharing the Animated-Factorization-Diagrams website and visualization with their secondary or middle-grades mathematics students. One extension of the lesson could include colorized factorization diagram posters (see a sample created by Yorgey [2012] on his website) to challenge teachers to consider how coloring based on the numbers (base 10) produces patterns in the resulting visualizations of the first 120 natural numbers arranged into columns of 10.

This is an integrated STEAM lesson because of the interplay between mathematics ideas (i.e., using the definition of prime number, describing patterns showing where prime numbers appear, and justifying reasons that these patterns exist) and non-mathematics ideas (visualization of a mathematics concept using a visual representation of dots to depict factors of numbers). In addition, the use of interactive web technology invites exploration and motivates curiosity for understanding the patterns observed, easily allowing teachers to check and verify whether the pattern holds for larger numbers.

4 STEAM Lesson 2: Pendulums

Unrelated to the Prime lesson, the Pendulum lesson is also intended for mathematics teachers and begins with their using an interactive website to explore phenomena. We developed this lesson by building on previously established work detailing how pendulums can be used to integrate mathematics and science (Brueningsen et al., 1997; McLeod et al., 2015; Wijaya & Widodo, 2021). This lesson has five parts, each taking about 15 minutes.

4.1 *Part 1: Exploring Quantities in a Swinging Pendulum*

To launch the activity, we asked teachers to explore an online interactive pendulum lab (Dubson & Rouinfar, 2021). The interactive pendulum lab has controllable features, including a ruler, stopwatch, period trace, and compass measuring the angle of the pendulum, length of the string, mass of the weight, gravity, friction, and time. The visualization of how each pendulum travels

helps learners to identify the functions of various factors. A double pendulum button appears at the bottom of the web page, but we asked teachers to explore only the single pendulum. After a few minutes of exploring, we asked teachers to answer the following questions in groups of three or four:
- State the question your group wishes to explore.
- Name the quantities involved in answering your question. When naming each quantity, include the units of the quantity and whether that quantity should vary or remain constant when answering your question.
- Describe what you could do to answer your question.

Common group responses involved how the period of the pendulum is affected by the length of the pendulum, the mass, the gravity, and the friction. Group descriptions of their experiment included what quantities they would vary and how they would record the relationship, but sometimes we pressed for more details. As groups finished answering these questions, we gave them instructions to move forward with their plan and record what they had found.

4.2 *Part 2: Exploring the Relationship between Period and Length of the Pendulum String*

We then shifted the focus to the relationship between the period of a pendulum and the length of the string. We asked teachers to keep all other quantities constant in their exploration: leave the default values for mass, gravity, and friction, and start each pendulum swing at 45 degrees. The following prompts were given to teachers to respond in small groups:
- Use the website to measure the period of a pendulum based on various string lengths, again keeping all other quantities constant.
    1. Create a table of values for the periods corresponding to different lengths of the pendulum.
    2. Be sure to have at least 10 data points that include a variety of lengths.
    3. Share your visual representation of the class data using a Google spreadsheet.

Actual class data resulting from these questions appear in Part 3.

4.3 *Part 3: Determining the Relationship between the Period and String Length*

Once all groups contributed to the class Google spreadsheet, we graphed the data in green (see Figure 11.1), including additional data in blue from in-class experiments completed in previous semesters, when we examined the period

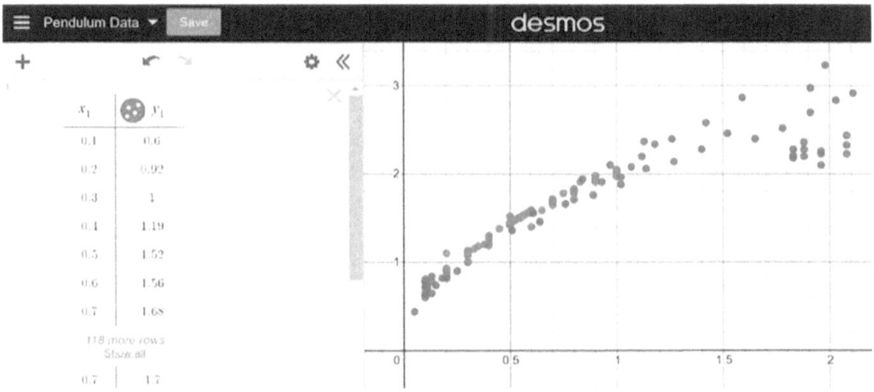

FIGURE 11.1 The Desmos graph of class data (green) with previous class data (blue), comparing the length of the string on the horizontal x-axis (measured in meters) to the period (measured in seconds)

with longer pendulum strings. We reminded teachers that the horizontal x-axis was the length of the string in meters and the vertical y-axis was the period measured in seconds.

We gave teachers the following questions to answer:

- Examine the following Desmos graph: https://www.desmos.com/calculator/z8uzgfrjs7
  1. How does our virtual data compare with the in-class data?
  2. What patterns do you notice when the string length is near 2 meters? Why might this happen? Hint: Visualize your group measuring the period with a 6-foot string.
- On the left side of the Desmos website, go to the empty input bar 3 and input *ax*. Click to add *a* as a slider. Then edit your function, testing out functions that might model the data. Use the slider for parameter *a* to quickly adjust the function you create to better model the data.

4.4 *Part 4: Understanding the Relationship between the Period and String Length*

We then had groups present the functions they created by displaying their Desmos graphs for all to see. In our experience, some groups always recognized that the square-root function accurately models the data (see Figure 11.2).

After the presentation, we led teachers to integrate science concepts and background information. We shared that physicists have found the same relationship between pendulum string length and period, which can be generalized as follows: let T represent the period in seconds; L, the string length in

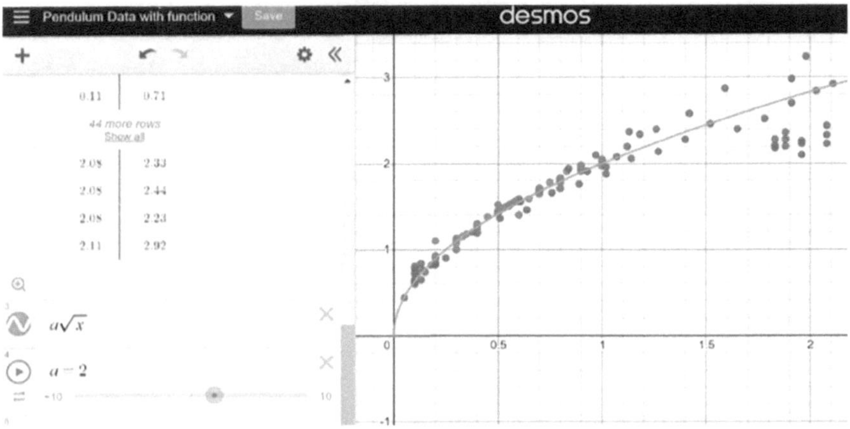

FIGURE 11.2 The Desmos graph of a function modeling the relationship between the pendulum string length and period

meters; g, the gravity in meters per second squared; and k, some constant of proportionality. Then the relationship can be modeled as follows:

$$T = 2\pi\sqrt{\frac{L}{9.8}}$$

To simplify, the relationship between the quantities can be modeled as follows:

$$T \approx 2.00709\sqrt{L}$$

This shows the accuracy of the model of $T = 2\sqrt{L}$ in Figure 11.2. We then had teachers unpack this information by discussing the following questions in small groups:

- Why does including the quantity of gravity in the equation make sense?
- How might you reason about the units of each quantity in this relationship and how they fit together in the equation?

Regarding the second question, ideas from dimensional analysis in physics supported teachers in understanding how the units of each quantity make sense in the equation (e.g., Burchat, 2007; Pavelich, 2011).

4.5 *Part 5: Creating Visual Patterns in Multiple Swinging Pendulums*

The final part of the activity began with the teachers watching "Pendulum Waves," a YouTube video (Harvard Natural Sciences Lecture Demonstrations, 2010) depicting 15 "uncoupled simple pendulums of monotonically increasing lengths danc[ing] together to produce visual traveling waves, standing waves, beating, and (seemingly) random motion." We asked teachers to determine how we could create this pendulum wave. We shared the information in the video description (Harvard Natural Sciences Lecture Demonstrations, 2010), which stated:

> The length of the longest pendulum has been adjusted so that it executes 51 oscillations in 60 seconds, and the length of each successive shorter pendulum is carefully adjusted so that it executes one additional oscillation in this period. Thus, the 15th pendulum (shortest) undergoes 65 oscillations in 60 seconds.

The overarching question teachers worked to solve was the following: How could we make such a pendulum wave? What do the lengths of each of the 15 pendulums need to be? Teachers created a list of pendulum lengths necessary to create the visual patterns seen in the video.

4.6 *Pendulum Lesson Discussion*

Pendulum waves have been discussed for some time in the area of physics (Berg, 1991; Flaten & Parendo, 2001), but we have not seen them used to motivate exploration of the way mathematical relationships create the visual patterns observed. The pendulum lesson requires learners to consider the quantities explored, how they are measured, and how they relate to one another, which corresponds to the Common Core State Standard (2010) Mathematical Practice #2: Reasoning quantitatively and abstractly (National Council of Teachers of Mathematics, 2014). In addition, recognizing and graphing the square root function appears in the Common Core high school mathematics standards (F-IF), and creating the pendulum wave requires learners to solve equations and address algebra standards (A-REI).

In terms of the visualization of data, guiding students to observe the pendulum movement (pendulum wave), associated with the way measurement could be calculated, helps them to understand abstract mathematic concepts, what Grodoski (2018) called "visualizing data" in art education. Both Presmeg (2006) and Sholihah and Maryono (2020) pointed out that visualization, "especially mental images of subjects controlling student thinking" (Sholihah & Maryono, 2020, p. 180), helps learners solve mathematical problems.

5 Conclusion

STEAM has the potential for a significant impact on the mathematics and mathematics education community, especially when using visual art to assist young learners in understanding abstract mathematics concepts. STEAM education can play a role in making mathematics more accessible and increasing integrated content representations within the discipline. By presenting a variety of visualized graphs, teachers can help students learn mathematics through visual aids that include multiple ways of visualizing data. With the inclusion of visual art into STEM, learners can not only visualize data but also "make visible relationships among [sets of] data" (Grodoski, 2018, p. 39). STEAM education efforts are expanding, with STEM schools placing increased focus on mathematical literacy and integrating content domains (Aguilera & Ortiz-Revilla, 2021). We hope this chapter provides teachers with instructional resources to consider implementing STEAM lessons in K–12 education.

References

Aguilera, D., & Ortiz-Revilla, J. (2021). STEM vs. STEAM education and student creativity: A systematic literature review. *Education Sciences, 11*(7), Article 331. https://doi.org/10.3390/educsci11070331

Autenrieth, R., Lewis, C., & Butler-Perry, K. (2017). Long-term impact of the enrichment experiences in engineering (E3) summer teacher program. *Journal of STEM Education, 18*(1), 25–31.

Berg, R. E. (1991). Pendulum waves: A demonstration of wave motion using pendula. *American Journal of Physics, 59*(2), 186–187. https://doi.org/10.1119/1.16608

Boice, K. L., Jackson, J. R., Alemdar, M., Rao, A. E., Grossman, S., & Usselman, M. (2021). Supporting teachers on their STEAM journey: A collaborative STEAM teacher training program. *Education Sciences, 11*(3), Article 105. https://doi.org/10.3390/educsci11030105

Brophy, S., Klein, S., Portsmore, M., & Roger, C. (2008). Advancing engineering education in P–12 classrooms. *Journal of Engineering Education, 97*(3), 369–387. https://doi.org/10.1002/j.2168-9830.2008.tb00985.x

Brueningsen, C., Brueningsen, E., & Bower, B. (1997). *Math and science in motion: Activities for middle school.* Texas Instruments Incorporated.

Burchat, P. (2007). *Lecture 2: Dimensional analysis - from biology to cosmology.* Stanford University. https://web.stanford.edu/~rpam/dropoff/Phys041N/lecture2_dimanalysis.pdf

Common Core State Standards Initiative. (2010). *Common core state standards for mathematics.* National Governors Association Center for Best Practices and the Council of Chief State School Officers. http://www.corestandards.org/wp-content/uploads/Math_ Standards.pdf

Dubson, M., & Rouinfar, A. (2021). *Pendulum lab.* University of Colorado Boulder. https://phet.colorado.edu/sims/html/pendulum-lab/latest/pendulum-lab_en.html

Flaten, J. A., & Parendo, K. A. (2001). Pendulum waves: A lesson in aliasing. *American Journal of Physics, 69*(7), 778–782. https://doi.org/10.1119/1.1349543

Glassmeyer, D., Hsieh, K., & Nguyen, L. (2018, December). *Integrating STEAM in the arts classroom: Examining student knowledge and teacher reflections* [Paper presentation]. The 2018 International Conference and Exhibition on Art and Design, National Pingtung University, Pingtung City, Taiwan.

Grodoski, C. (2018). Visualizing data in art education: Research, practice, and advocacy. *Art Education, 71*(2), 36–45. https://doi.org/10.1080/00043125.2018.1414537

Harvard Natural Sciences Lecture Demonstrations. (2010, June 9). *Pendulum waves* [Video]. YouTube. https://youtu.be/yVkdfJ9PkRQ

Henriksen, D., Mehta, R., & Mehta, S. (2019). Design thinking gives STEAM to teaching: A framework that breaks disciplinary boundaries. In M. S. Khine & S. Areepattamannil (Eds.), *STEAM education* (pp. 57–78). Springer.

Hsieh, K., Glassmeyer, D., & Nguyen, L. (2019). The collaborative teaching and learning: High school STEAM project-based learning and lantern design. *International Journal of Arts Education, 17*(1), 33–67.

McLeod, K., Steele, M., & Brown, S. (2015, February). *Design features and outcomes in common core professional development for high school: Functions and modeling* [Paper presentation]. Association of Mathematics Teacher Educators Annual Conference, Orlando, FL, United States.

National Art Education Association. (2022, June 1). *NAEA position statement on STEAM education*. https://www.arteducators.org/advocacy-policy/articles/552-naea-position-statement-on-steam-education

National Council of Teachers of Mathematics. (2014). *Principles to actions: Ensuring mathematical success for all*. Author.

National Science Board. (2022). *Science and engineering indicators 2022*. National Science Foundation Report 430. https://www.nsf.gov/statistics/indicators/

Neale, V. (2017). *Closing the gap: The quest to understand prime numbers*. Oxford University Press.

Nutov, L. (2021). Integrating visual arts into the mathematics curriculum: The case of pre-service teachers. *Teaching and Teacher Education, 97*, 1–11. http://dx.doi.org/10.1016/j.tate.2020.103218

Pavelich, L. (2011, September 23). Fun with dimensional analysis 1 – Simple pendulum. *Scrub Physics*. https://leepavelich.wordpress.com/2011/09/23/fun-with-dimensional-analysis-1-simple-pendulum/

Perignat, E., & Katz-Buonincontro, J. (2019). STEAM in practice and research: An integrative literature review. *Thinking Skills and Creativity, 31*, 31–43. https://psycnet.apa.org/doi/10.1016/j.tsc.2018.10.002

Presmeg, N. (2006). Research on visualization in learning and teaching mathematics. In A. Gutierrez & P. Boero (Eds.), *Handbook of research on the psychology of mathematics education: Past, present and future* (pp. 205–236). Sense.

Sholihah, U., & Maryono, M. (2020). Students' visual thinking ability in solving the integral problem. *Journal of Research and Advances in Mathematics Education, 5*(2), 175–186.

Suganda, E., Latifah, S., Irwandani, I., Sari, P. M., Rahmayanti, H., Ichsan, I. Z., & Rahman, M. M. (2021). STEAM and environment on students' creative-thinking skills: A meta-analysis study. *Journal of Physics: Conference Series 1796*(1), 1–9. http://dx.doi.org/10.1088/1742-6596/1796/1/012101

Tayebi, A., Gómez, J., & Delgado, C. (2021). Analysis on the lack of motivation and dropout in engineering students in Spain. *IEEE Access*, *9*, 66253–66265. http://dx.doi.org/10.1109/ACCESS.2021.3076751

Von Worley, S. (Oct. 5, 2012). *Dance, factors, dance: A variation on [Brent] Yorgey's factorization diagrams*. http://www.datapointed.net/visualizations/math/factorization/animated-diagrams/

Wijaya, P. A., & Widodo, A. (2021, March). Virtual experiment of simple pendulum to improve students' conceptual understanding. *Journal of Physics: Conference Series, 1806*(1), Article 012133. http://dx.doi.org/10.1088/1742-6596/1806/1/012133

Yorgey, B. (2012, Oct. 5). Factorization diagrams. *The math less traveled*. https://mathlesstraveled.com/2012/10/05/factorization-diagrams/

CHAPTER 12

STEAM in Higher Education Art Programs

Snow Yunxue Fu and Luke Hampton

Abstract

Using the emerging 3D-technology curriculum developed by Snow Yunxue Fu of the Department of Photography and Imaging at the Tisch School of the Arts at New York University as an exemplar of a technology-based experimental art program in higher education, where STEAM strategies thrive, the authors propose that higher education programs can inspire K–12 educators seeking to develop STEAM curricula. The case study focuses on a central class project in which Fu incorporates cutting-edge virtual reality, Web 3.0 technologies, and collaborative learning strategies to prepare her students for the future.

Keywords

STEAM – higher education – K–12 – emerging media technology – new media

1 A Legacy of Technology and Innovation

On an October evening in 1966 in Manhattan's 69th Regiment Armory, the renowned artist Robert Rauschenberg met with 30 Bell Laboratory electronics engineers and 10 performance artists to demonstrate the fruits of their collaborations before an audience of engineers, artists, and art critics. The performances at the Armory continued over the course of nine evenings, hence the event gained its colloquial title—*9 Evenings*. During the preceding 10 months, the cohort of engineers and artists had collaborated under the supervision of Rauschenberg, an artist known for his interdisciplinary, collaborative, and experimental artmaking, their goal to manifest theatrical performances by pushing the boundaries of the application of technology in artistic and performative contexts (Hultén & Königsberg, 1966).

In an extraordinary display of problem-solving and technical innovation, the collaborations produced unprecedented theatrical uses of cutting-edge technology, including large-scale closed-circuit, fiber-optic, and infrared

 | DOI:10.1163/9789004714748_013

camera projections; a Doppler sonar device that translated dance movements into sound; portable wireless FM transmitters and amplifiers; and wheeled remote-controlled stage platforms. Like many of Rauschenberg's endeavors, such a display had never been seen before. In 1966 computer electronic engineering and the arts were largely unacquainted, yet those nine evenings captured a spirit of cross-disciplinary comradery and innovation that can be remembered with admiration. Despite the unprecedented and inspirational nature of the performances, however, they initially received largely unpleasant responses from their audience (Hultén & Königsberg, 1966).

After the events took place, critics reported that the film and photographic documentation of the evenings captured the spirit, vision, and potential of such collaborations, but *9 Evenings* was a chaotic and inefficient symphony of short-circuits and setbacks, often with hours between performances due to malfunctioning set pieces. Brian O'Doherty (1966/2006), a prominent art critic in attendance, mused, "These records [of *9 Evenings*] tend to be more impressive imaginatively than the events themselves; the historical audience … will regret they were not there. The actual audience often regretted that they were." This is the nature of experimentation when artistic vision and technical limitations meet; however, their efforts were hardly in vain. The setbacks and limitations of the technology at the time were no less a catalyst for artistic and technological innovations. O'Doherty later stated that "the impact of the affair has been enormous" and only after piecing together the broader impact can we begin to qualify what actually took place (p. 79).

2 Changing Times

The tale of *9 Evenings*—with its perceived successes and failures—reveals the value and challenges of collaborating across disciplines. The initial responses of the stakeholders in attendance imply the differences in expectations regarding what fields like science, technology, engineering, mathematics, and the arts should be producing and how they should produce it. These expectations stem in part from the differences in the arts and sciences as well as math and engineering and the processes of generating ideas, innovating, and problem-solving in each. The sciences are often associated with efficiency, effectiveness, and reliable data while the arts prioritize creativity, self-expression, and activism; nevertheless, cross-disciplinary thinking has become more important in increasingly technologically integrated societies, where technology is a conduit for the collaboration of the sciences, mathematics, engineering, and the arts.

Technological advancements like artificial intelligence (AI), Web 3.0, and the metaverse (Ball, 2021) are primed to transform job markets in the not-too-distant future. These changes will likely shift hiring trends toward soft skills like creative problem-solving and collaboration. Today, cross-disciplinary skill sets are widely regarded as an advantage. In modern STEM industries, such as computer engineering, entertainment, and cyber security, candidates are increasingly expected to have varied skill sets across creative, collaborative, and technological literacies (Young, 2019). Consequently, national expectations regarding the skills that higher education and K–12 school systems should teach have shifted as well. This shift in general is one leaning toward interdisciplinarity, which in education generally applies to four realms—knowledge, research, education, and theory—combining components of two or more disciplines in the search for or creation of new knowledge or artistic expression (Nissani, 1995). This application of cross-disciplinary thinking intended by STEAM advocates, where the differences in research methodologies and systems of thought are applied to solving complex problems, does not often fit neatly into one subject (Sudderth, 2023).

Proactive education models, such as STEAM, STREAM (Debroy, 2017), and STREAMSS (Brown & Bousalis, 2018) are on the rise. The addition of A for art, R for reading and writing, and SS for social studies reflect an intention—beyond alphabet soup—to integrate the thinking systems, problem-solving strategies, and knowledge production of various academic disciplines to bolster the efficacy of traditional STEM models of education. New educational models gain traction by proclaiming lofty inspirational outcomes, often with questions remaining as to how interdisciplinary synergy can occur on the ground, in classrooms, and in lesson plans. Despite this, the STEAM model has emerged as one of the most well-known movements, with applications that span from K–12 to higher education.

3 Building STEAM

The STEAM acronym represents a model that integrates the arts into the traditional STEM model. In this relatively new model, nailing down precisely what implementing STEAM looks like in a classroom can be elusive, as the implementation can be significantly influenced by which letter in the acronym the educator's or author's background is based. For example, scientists may emphasize art integration as data visualization skills (Hepworth & Canon, 2018), those in technology and engineering may incorporate iterative design thinking processes (Hare, 2016), and educators who teach highly diverse classrooms may emphasize STEAM's capacity to create equitable and positive encounters with STEM content earlier in the academic journey of student demographics who historically engage in STEM less, such as female and minority students (Compuchild, 2023).

Variations in perspective are both the strength and weakness of education models like STEAM, which aim to integrate often contrasting viewpoints on how best to interpret the value of learning outcomes. Fortunately, advocates of STEAM aim not for a superficial adoption of artspeak within STEM lesson plans but a substantial interdisciplinary, philosophical, and practical integration. The potential for synergy is apparent in that most disciplines "involve [the] creative process, and none use[s] just one method of inquiry and investigation" (Lathan, n.d.). Although current literature and discourse on the STEAM model may lack a conclusive and widely circulated definition or template for implementation, a set of central objectives and principles can be summarized as a focus on real-world problem solving through interdisciplinary creativity, collaboration, and innovation. The benefit proposed by STEAM then and interdisciplinarity in general is its capacity to develop adaptable nimble young minds.

From an economic standpoint, what these principles and capacities ultimately translate to is college and career readiness, which have long been one of the main filters through which education initiatives are assessed. STEM education began, in part, as a foundation to encourage students in the United States to study and later work in STEM fields and by extension to secure their place at the top of global academic scores and as innovators in the workforce. If STEM education was in large part intended to prepare students for future careers, a significant alteration to STEM education implies that expected trends in future careers have also changed.

Despite a mountain of research long validating the inclusion of the arts in school systems, the periodic pendulum swinging of government funding and cutbacks have too often labeled them expendable. "The addition of [the arts in STEM] is important in part because practices such as modeling, developing explanations, and engaging in critique and evaluation have too often been underemphasized in the context of math and science education" (Lathan, n.d.). The addition of the A in STEM implies that the strategies of traditional STEM K–12 education are no longer as apt at producing the skill sets required for future graduates and that policy makers more readily see the value of the arts in K–12 education. As industry tasks are replaced by AI systems, job recruiters will look for skill sets that AI cannot replace, including developing abstract concepts, creative problem solving, collaboration across disciplines, and moral awareness (Chowdhury, 2022; Stahl, 2021). "Education is under pressure to respond to a changing world. As repetitive tasks are eroded by technology and outsourcing, the ability to solve novel problems has become increasingly vital" (Vasagar, 2014). It is no coincidence, then, that the skills reflected in future predictions parallel the learning outcomes represented in the STEAM education model.

The future-focused competencies with which STEAM concerns itself represent an ecosystem of influence, where the push–pull influence of job markets and curriculum reform are visible throughout K–20 education. The catalyst of changes in these major markets often begins as developments in technologies that have the potential to transform STEM and creative industries. This influence trickles down through higher education and into K–12 policy and curricula that aim to preemptively adapt. Educators who wish to implement initiatives like STEAM in their classrooms must likewise become adaptive, keeping abreast of how new technologies shape STEM fields and which education strategies to apply in their individual contexts.

Advocates of STEAM can draw inspiration from higher education programs that combine art and technology to develop strategies and curricula for implementation. Like the electronic engineers and artists of *9 Evenings*, experts in such higher education arts and technology programs often experiment with cutting-edge technologies, spurring advancements in artistic expression and practical technological applications. In time these cutting-edge innovations will become commonplace. These opportunities for experimentation in large part come from teaching artists like Rauschenberg who work as professionals in the arts while teaching in higher education.

STEAM itself was initially incubated on the fertile grounds of a prominent higher education art school (Rhode Island School of Design [RISD], 2014). Well-known among arts professionals, educators, and students for blurring the lines between engineering, design, and aesthetics, the Rhode Island School of Design (RISD) concerns itself with integrating experimentation, technology, and industry. STEAM, as a model, mirrors this intermingling. Interim RISD President Rosanne Somerson affirmed in a 2013 Capitol Hill address at the launching of the Congressional STEAM Caucus that "RISD is proud of our STEAM initiative's continued momentum and growth, and its development into a broad national conversation on educational innovation and global competitiveness" (RISD, 2014). Congressman Aaron Schock concurred, "The jobs of the future will require far more than mere proficiency in these areas. STEAM harnesses the synergy between the arts and STEM to produce graduates with the skills industries identify as vital in new hires, including collaboration, trial and error, divergent thinking skills, and dynamic problem solving" (RISD, 2014).

4 Emerging Media Curricula in Higher Education

In the following section we explore the current applications of interdisciplinary strategies through a case study focusing on teaching artist Snow Yunxue Fu and the curriculum she developed at New York University (NYU). We highlight Snow's practice as an artist and her seminal virtual reality (VR) project, which she has incorporated into her courses, in all of which the future-focused approaches between fields intersect and from which K–12 educators can draw insights.

Snow is an assistant arts professor in the Department of Photography and Imaging (DPI) at the Tisch School of the Arts at NYU. The department offers a four-year B.F.A. program with opportunities for students to work within a breadth of analog and digital photo-based image making, including multimedia, new media, immersive, and 3D simulation technologies (NYU Tisch, n.d.). Snow was hired in 2019 to expand the DPI curriculum with new media courses, helping to ensure the program remains current with growing trends within imaging theory, contemporary art practices, and creative industries. She has integrated new courses into the program while supporting the broader vision of Tisch's Institute of Emerging Media, the future-focused multidepartmental initiative under which DPI operates. As a university faculty member in the field of imaging and new media art, Snow shares the same convictions as many primary and secondary educators: to equip new generations of students with the knowledge and skills to navigate a changing world. She has developed courses that funnel the interdisciplinary skill sets and collaborative experiences students will need to compete in a world that will look different from the one they entered as freshman.

The development of these convictions is rooted in her prior studies as an artist in interdisciplinary art programs and an 11-year art practice focusing on emerging 3D digital technologies. Coming from an extensive multidisciplinary background, her artwork focuses on experimentation with emerging media technologies, such as VR, augmented reality (AR), and Web 3.0 in its capacity to engage VR and AR experiences remotely. Her media of choice are industry software standards, including Maya, Z-Brush, RealFlow, Poser, Unity, and Unreal Engine, culminating in 3D image stills, animation, moving images, media installations, and interactive projects. She notes that at first glance 3D media technology may imply a conceptual detachment from photography. For Snow, however, 3D digital media and the spaces within virtual reality share in and continue the histories and practices of filmmaking, photography, painting, sculpture, theatre, architecture, Earth art, and interactive arts.

Her work merges sociological, anthropological, and philosophical explorations into the techno sublime—a technologically simulated experience with the sublime. Her VR experiences transport participants to open-world landscapes, where they can explore ethereal yet familiar environments. The exact scale is uncertain, and distortions are often caused by interacting with objects and surroundings. Her environments are based on natural wonders, such as oceanic trenches, caves, and glaciers, each frequently entangled with abstracted manmade artifacts, including cityscapes, wildfires, and debris. Depending on the artwork, participants can virtually interact with the environments from a first-person viewpoint or over the shoulder of an avatar (See Figure 12.1). Conceptually, her VR work echoes the investigations of historical Western and Chinese painters who peered into the significance of our capacity to experience the sublime in nature. Through VR she creates spaces for participants to explore the implications of this capacity in a virtual space, where they are offered an opportunity to examine the implications of being placed at the threshold of their perception within the VR experience.

5 Becoming a Teaching Artist in New Media Art

Snow's involvement with 3D computer graphics chronologically parallels the same histories of interdisciplinary art programs that spawned the principles leading to RISD's articulation of STEAM. She was introduced to computer graphics software in the early 2000s, not long after a time when 3D computer graphics were more readily associated with mainstream entertainment and

FIGURE 12.1 Screenshot of an avatar running through an abstracted city scape. Fu, Y. (n.d.). *Run – Valley City Arts Game Recording Building Level (Take2), 2021* (from https://snowyunxuefu.com/section/505997-Run%20-%20Valley%20City.html)

novelty, such as in commercial animation, cinematic special effects, and gaming. By that time 3D digital media content had gained traction in art and design school curricula, usually through teaching artists who had experimented with 3D work for decades, often out of the limelight of the broader art world. In Snow's graduate studies at the School of the Art Institute of Chicago (SAIC), in 2013, she was fortunate to take a course with one such teaching artist, Claudia Hart, a pioneer in teaching 3D software purely as an art medium (Hart, n.d.). Like others in the growing new media community, Snow found in 3D an avenue to fully express her practice, one that reflected the changes in society related to technology and the social commentary she sought to engage.

At the time of Snow's graduate studies, schools like SAIC offered courses not only across media but across departments and facilities. Introductions to new media and 3D computer imaging coincided with a push in higher education art schools toward interdisciplinary programming. In this openness to explore at SAIC, Snow found a method of learning, process, and production that matched the way she learned naturally. Her introduction to 3D software and new media art now mirrors that of many of her students at DPI, who often come to the program from a background seemingly unrelated to new media and then discover in 3D imaging a new avenue of expression and theory to explore.

In collaboration with the chair of DPI and the Tisch administration, Snow developed a series of four courses focusing on digital media with a new media and postphotographic framework: MultiMedia Intro to Post-Photographic Imaging, Emerging Media Studio Digital Bodies, Emerging Media Studio Fluid Environments, and the sci-fi digital SFX inspired course, Future Tense. Postphotographic here does not mean nonphotographic. At its simplest, postphotography is a relatively new theory and conceptual approach, a reevaluation of how current societies relate to images and their creation, often in context with the technologies we use to engage and create those images. Exploring definitions of postphotography, photographer Joan Fontcuberta (2015) wrote,

> We are interested in post-photography because it invites us to remove thought from our actions in relation to the image, launching us on a philosophical exercise that engages with the experience of our digital life… Post-Photography is not a style, movement or historical period. It is of interest here due primarily to the opportunity that it offers us to clarify what photography is, and who we are, today. In other words, it offers us an opportunity to re-examine photography and re-analyze our nature as subjects. (p.12)

Likewise, Snow's courses emphasize multidisciplinary creative practices, particularly involving the fields of 3D imaging and digital simulation, and—like her art practice—an engagement with emerging online technologies and the philosophies and social commentary that resonate from their use.

A highlight of developing her curriculum has been creating projects and opportunities for students to create work that exhibits in the metaverse—not to be confused with Mark Zuckerberg's annexation of the term by his company Meta. The metaverse at its simplest is a growing network of online VR spaces where users can interact virtually in real-time via VR headsets, usually as self-styled avatars. These VR spaces are commonly hosted by online platforms specifically designed to accommodate the taxing hardware and requirements of VR experiences. Ball (2021) explained in more detail: "The Metaverse is an expansive network of persistent, real-time rendered 3D worlds and simulations that support continuity of identity, objects, history, payments, and entitlements, and can be experienced synchronously by an effectively unlimited number of users, each with an individual sense of presence."

6 Interdisciplinary Learning Strategies

Snow experiments with emerging technologies in her art practice and incorporates them into her course projects. The technical skills embedded in these projects are likely unfamiliar to the uninitiated, including character rigging, motion capture, object modeling, and XYZ axis manipulation, but they can be summarized broadly as 3D software tools. Such tools allow for the manipulation of uploaded data or a created virtual product. Similar to Photoshop, some tools can be used by novices, for example, photo editing sliders; others require exceptional expertise, an awareness of the software's interconnectedness, and the manipulation of raw data. Snow's courses vary in technical skill levels, allowing for progression over several projects throughout the semester, shepherding student learning toward a foundation of skills, a final project, and a student exhibition.

Snow sees the most important of these foundational skills and experiences as the collaboration that occurs among peers, applications of digital technologies in the arts, and the analysis of concepts and systems of thought from guest speakers, presentations, and online tutorials—all of which must be thoroughly integrated conceptually by students during student presentations and critiques. The conceptual and technical integration is produced through a continuum of opportunities, first inspired, then self-directed; and while that self-direction plays out, students are allowed to collaborate as they encounter problems, sharing what they have learned in the process.

Inspiration for projects often comes through class discussions, critiques, lectures, tutorials, and guest speakers arranged through Snow's network of artists and professionals. A student may be inspired to use a software tool, a third-party software, or external hardware in a way that was not specifically taught. If the intentions of the students are to further their learning, to enhance their final projects, and to share what they learned in class, Snow encourages those intentions.

The hype and anticipation surrounding the metaverse notwithstanding, the technologies, online infrastructure, ethics, and philosophy of the metaverse are far from fleshed out. This is apparent in the recent floundering of Meta as Zuckerberg continues to invest in it despite minimal user engagement and technical setbacks (Rapoza, 2022). These limitations provide ripe conditions in classrooms for problem solving across media, art forms, and disciplines to arrive at solutions mirrored in real-world industries and art practices. Given the interconnectedness of software user interfaces, when a student experiments with one, the possibilities for cross-platform application grows exponentially as students discuss and collaborate. This creates a fluidity within Snow's courses that can at times radically change the parameters she originally set, also allowing for students to push past their preconceived notions of what photography or new media can be within the context of a classroom. Such pedagogical strategies make room for students to enjoy and become aware of their own learning.

7 A Seminal Class Project

One ongoing project has encapsulated the objectives represented in Snow's curriculum and classroom pedagogy. From 2021 to 2023, Snow hosted student exhibitions virtually in the metaverse, where students, professors, alumni, family, and friends were able to join the exhibition. Through this process, students engaged with the metaverse and VR technologies. In summer 2021 Snow entered a partnership with the Paris-based DSLCollection (DSL) run by the entrepreneurial duo Sylvain and Dominique Levy alongside their daughter Karen Levy, who assists in globally building awareness of the collection. DSL was created in 2005, focusing on contemporary art made by Chinese artists; the collection comprises major works of 250 of them. DSL has distinguished itself from other collections through its philosophy: a belief that the 21st-century collector must think beyond established collection boundaries, particularly those related to the use of virtual exhibition technologies (Ho, 2018). The Levys explore the use of interactive virtual programs to help provide the means to share the experience of contemporary culture with a wider audience of artists, students, educators, and institutions, and invest in the development of metaverse art spaces to facilitate dialogue and collaboration across diverse fields in academic and professional contexts.

Snow was contracted by DSL to develop a virtual space for such facilitation in collaboration with her DPI curriculum and the work created by her students at NYU. This was to be a shared virtual space for exhibition, artist talks, workshops, and student organized events. With the help of her assistants, Phil Mulliken, Kamile Eidukeviciute, and Hailing Liu, she designed a virtual space modeled after Washington Square Park (WSP) in Manhattan's Greenwich Village. The virtual park site was built using the industry standard Maya 3D software and imported into the social VR platform Sansar, which represents a growing number of options for users to build their own 3D virtual worlds, serving as spaces to engage a wide array of communities with access to VR technology. Platforms like Sansar were predicated on early online community platforms like Second Life, which were popular in the early 2000s but lacked an immersive VR component. More modern platforms like Sansar are to Web 3.0 as Second Life was to Web 2.0 in that Sansar is fully VR compatible.

The WSP virtual space has taken on the energy of the physical site, paralleling the community engagement of the latter as a recreational and cultural hub for the NYU community and Manhattanites alike with many musicians, performers, and artists using the physical park daily. Reconstructed to the scale of the avatars, it is its own unique space within the aesthetics of the software and artistic inaccuracies (see Figure 12.2). The park's iconic arch is itself often seen as a symbol of NYU. The virtual space also references the histories of the park and offers indirect historical and contemporary societal commentary, most notably that in the 1800s the location of WSP was a burial ground for unidentified indigenous people, those who were enslaved, those living in poverty, and in the 1900s many of those who fell to the yellow fever epidemic (*The Skeletons*, 2021). The current vibrance, community, and histories of the park reflect not only the progress that has been made over time in equality but also the inequalities, such as those brought to light in the diverse neighborhoods of New York City during the COVID-19 pandemic. The park represents, by its name, the foundations of the country and higher education in America and the future potential that it holds. The face of the VR WSPark project is positioned toward an enthusiasm and hope for that future, where virtual spaces have the potential to become an equalizing factor.

The WSP virtual space has been used for multiple events to date, debuting in late 2021 under the name, VR WSPark. The first event hosted in the virtual space was titled *Land of Post Photo*, an exhibition of the work of 10 of Snow's former NYU students presented on large billboard structures placed throughout the

FIGURE 12.2 Virtual aerial view of VR WSPark. Fu, Y. (n.d.). *VR WSPark metaverse project* (from https://snowyunxuefu.com/artwork/5087411ArtLabb%20Opening%20on%20VR%20WSPark.html)

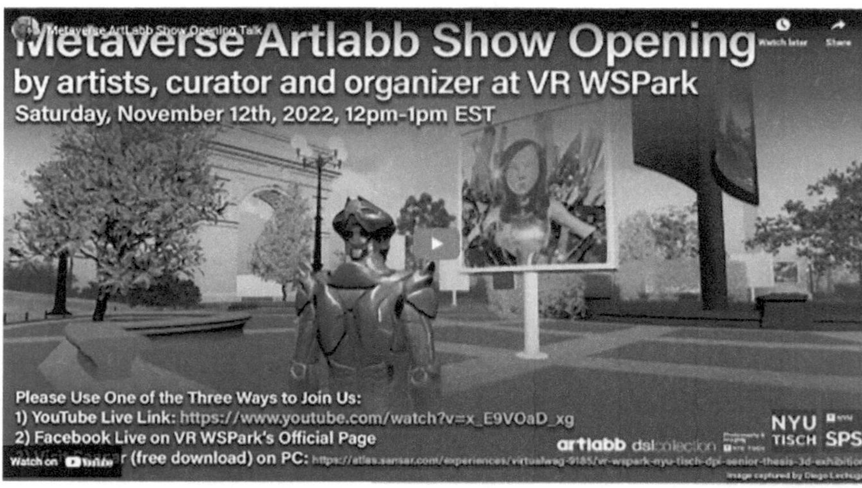

FIGURE 12.3 YouTube thumbnail advertising the online Metaverse Artlabb Show Opening. Fu, Y. (n.d.). *VR WSPark metaverse project* (from https://snowyunxuefu.com/artwork/5087411ArtLabb%20Opening%20on%20VR%20WSPark.html)

virtual park. For the opening, current students, alumni, invited guests, and the public could participate by logging on as avatars or joining through the livestream, which was directly uploaded to YouTube. The students and alumni discussed their works during a scheduled artist talk where participants could ask questions and view the artwork at their own pace (see Figure 12.3).

8 Reflections and Takeaways

8.1 *Self-Driven and Collaborative Learning*

The effectiveness of Snow's experimental classroom learning environment hinges on three previously suggested elements: individualized student-led projects, collaborative problem solving, and the expertise of the instructor as a guide. Snow sets the parameters of a given project, yet students can take the conceptual and technical aspects in any direction they choose, being allowed to use multiple digital tools and techniques at any point during the duration of the project. With this freedom, multiple problems often arise simultaneously. Students are expected to learn by doing through trial and error. If problems occur, or outright roadblocks to moving forward, they are expected to collaborate with their peers to discuss solutions and locate learning resources. As the problem increases in complexity and is shared by the class as a group, Snow incorporates new processes into the lessons to find solutions. One of her students reflected on this style of learning:

> I used to think a topic or theme has to be set before we start working on it …; however, in recent years, I learned that sometimes the most in-depth notion of the project emerges within the making process, not what I initially set it …. I find this mindset very rewarding and profound for more creative projects. (personal communication, Nov. 13, 2022)

Although the expectations for students to be self-directed and self-driven are naturally higher in higher education, self-directed projects can be applied in any K–12 classroom and have become increasingly popular in K–12 art curricula, although applications of the concept and the benefits therein are possible in most subjects.

8.2 *User Interface Literacy*

Having taught students of a wide range of ages, Snow has observed that each generation is ever more technologically literate, particularly the youngest students—digital natives born with phones in their hand. She leans on that native literacy to quickly bridge gaps in technical skills in her classrooms, which are often filled with a diverse mixture of students from first years to fourth years, international to domestic, and digital to analog focused. Part of native digital literacy means that most students have an intuitive understanding of the language of user interfaces at least by the time they enter higher education. Lesson plans intended to introduce software take much less time compared to the more methodical step-by-step process that would have been expected even a decade ago. K–12 students can and should be introduced to as many software and user interfaces as possible if they wish to build this literacy. Experimenting with user interfaces can be a project on its own, even for common software like the Microsoft Suite. Students can approach their learning through a set time for experimentation, as well as opening opportunities for students to collaborate with their peers, to discuss, solve problems, and demonstrate what they learned. Fundamentally, these types of lessons will likely remain much longer with the students than a step-by-step print-out from a software manual.

8.3 *Funding and Resources*

For educators or administrators seeking to integrate innovative technologies into their schools, one of the largest and most common obstacles is funding for the equipment and software licenses necessary to run a program with contemporary capabilities; however, the technical requirements for emerging technologies, such as 3D and VR experiences, are scalable in today's world of online resources, powerful personal laptops, tablets, and smartphones. Nearly any phone with internet capabilities can run AR and VR applications, and educational programming resources are increasingly available and targeted at engaging K–12 students in interactive media and coding tutorials.

Beyond the classroom, Snow emphasizes the importance of building a network of support across institutions and local communities. Governmental and foundation funding has increased for technology-oriented education initiatives. Tech education-related nonprofits are on the rise, providing students in underfunded schools opportunities to interact with such equipment. For example, this support could include a network of like-minded local artists and educators who can pool resources and expertise to host events or create projects, both for students and teacher training. This network lends itself to creating opportunities for students to present their projects in public exhibitions and engage in real-world collaborations.

9 Tomorrow Together

As early as the 1960s, cornerstone artists like Robert Rauschenberg and his cohort of collaborating computer engineers and performance artists began the hard but exhilarating work of building vision between what were considered opposing systems of thought and expression. Later technophile artists, such as Nam June Paik, created artwork throughout the 1970s and onward that approached audiovisual technology as art making mediums (Hanhardt & Hakuta, 2012). Artists like Claudia Hart would eventually pioneer the expansion of interdisciplinary new media courses in higher education art programs.

Their creative use of digital media created opportunities for society to consider the implications of our relationship with emerging technology while creating curricula that prepared students to engage with and anticipate how such technology would change their future. Contemporary new media artists like Snow continue in that trajectory, exploring new theories, such as postphotography, and the web-based innovations that make them possible.

Embedded in the interdisciplinary approach of STEAM and new media curricula is a step away from traditional modes of education and high stakes standardized testing and a step toward the value of process, learning to learn, and creative problem solving. Therein lies one of the most appealing aspirations inherent in STEAM—that educators from many disciplines can work together to implement and experiment with self-directed, art-based projects that capitalize on innovation and creativity, incorporating the skills necessary in the 21st-century workplace (Buczynski et al., 2012). The promise of web 3.0 and the metaverse is a future of the seamless integration of our physical and digital selves. Society is now in a liminal space, where social media platforms and digital media have expanded to an unprecedented scope and reach, likely providing an array of benefits and unforeseen consequences. Within this liminal space, K–12 educators can draw insights and strategies from higher education technology-focused programs. When asked to describe the impact of Snow's courses, one of her students summarized: "We are exploring and understanding the oil brushes of today, and through this experience … we are inventing our tomorrow together" (personal communication, Nov 11, 2022).

References

Ball, M. (2021, June). Framework for the metaverse [Foreword to *The metaverse primer*]. https://www.matthewball.vc/all/forwardtothemetaverseprimer

Brown, S., & Bousalis, R. (2018). *Curriculum integration in contemporary teaching practice: Emerging research and opportunities.* IGI Global.

Buczynski, S., Ireland, K., Reed, S., & Lacanienta, E. (2012). Communicating science concepts through art: 21st-century skills in practice. *Science Scope, 35*(9), 29–35 (EJ996843). ERIC.

Chowdhury, M. (2022, May 27). *8 skills that can't be replaced by robots and automation. Analytics Insight.* https://www.analyticsinsight.net/8-skills-that-cant-be-replaced-by-robots-and-automation/

Compuchild. (2023). *Inclusion and equality through STEAM programs for kids.* https://compuchild.com/blog/inclusion-and-equality-through-steam-programs-for-kids

Debroy, A. (2017, October 7). What is STREAM education & why is It gaining popularity? *EdTechReview*. https://edtechreview.in/trends-insights/insights/2968-what-is-stream-education

Fu, Y. (n.d.). *Snow Yunxue Fu*. https://snowyunxuefu.com/home.html

Fontcuberta, J. (Ed.). (2015). *The post-photographic condition: Le Mois de la Photo a Montreal, international biennial of the contemporary image*. Kerber.

Hanhardt, J. G., & Hakuta, K. (2012). *Nam June Paik: Global visionary*. Smithsonian American Art Museum.

Hare, T. (2016, January 18). *Turn STEM to STEAM with the design thinking process*. The Art of Education University. https://theartofeducation.edu/2016/01/18/turn-stem-to-steam-with-the-design-thinking-process/

Hart, C. (n.d.). *Claudia Hart*. https://claudiahart.com/

Hepworth, K. J., & Canon, C. (2018). Improving science students' data visualizations: A STEAM-based approach. *Dialectic*, 2(1), 48–78. https://doi.org/10.3998/dialectic.14932326.0002.104

Ho, H. (2018). *DSLCollection: A collection of Chinese contemporary art*. https://www.dslbook.com/dslbook/

Hultén, P., & Königsberg, F. (Eds.). (1966). *9 Evenings: Theatre and engineering* [Exhibition program]. The Foundation for Contemporary Performance Arts, Inc. https://monoskop.org/images/5/5d/9_Evenings_Theater_and_Engineering_1966.pdf

Lathan, J. (n. d.). *Why STEAM is so important to 21st century education*. University of San Diego online. https://onlinedegrees.sandiego.edu/steam-education-in-schools/#STEAM

Nissani, M. (1995). Fruits, salads, and smoothies: A working definition of interdisciplinarity. *The Journal of Educational Thought*, 29(2), 121–128. https://doi.org/10.11575/jet.v29i2.52385

NYU Tisch. (n.d.). *Photography & Imaging*. https://tisch.nyu.edu/photo

O'Doherty, B. (2006). New York: 9 armored nights. In C. Morris (Ed.). 9 evenings *reconsidered: Art, theatre, and engineering, 1966* (pp. 79–81). MIT List Visual Arts Center. (Original essay published 1966)

Rapoza, K. (2022, October 21). The metaverse is failing, but this is one investment that will not die. *Forbes*. https://www.forbes.com/sites/kenrapoza/2022/10/21/the-metaverse-is-failing-but-this-is-one-investment-that-will-not-die/

Rhode Island School of Design. (2014, May 7). *Rhode Island School of Design launches STEAM map to demonstrate global activity and support for the movement*. https://www.risd.edu/news/for-press/press-releases/rhode-island-school-design-launches-steam-map-demonstrate-global

Stahl, A. (2021, March 10). How AI will impact the future of work and life. *Forbes*. https://www.forbes.com/sites/ashleystahl/2021/03/10/how-ai-will-impact-the-future-of-work-and-life/

Sudderth, A. (n.d.). A guide for interdisciplinary teaching and learning. *Rethink together*. https://xqsuperschool.org/rethinktogether/interdisciplinary-teaching-and-learning/

The skeletons of Washington Square. (October 12, 2021). *Washington Square Park Blog*. https://www.washingtonsquareparkblog.com/2021/10/12/skeletons-washington-square/

Vasagar, J. (2014, May 19). Countries that excel at problem-solving encourage critical thinking. *Financial Times*. https://www.ft.com/content/e512db9c-c643-11e3-ba0e-00144feabdc0

Young, K. (2019, July 3). Why creativity is important for all jobs, even in STEM. *Silicon Republic*. https://www.siliconrepublic.com/advice/creativity-jobs-stem-automation-ai

Glossary

3D simulation can refer to a singular computer graphics-generated virtual object or scenery, or—more broadly—as the field and practice of using 3D simulation technology

Augmented reality The real-time overlaying of interactive computer-generated graphics onto a viewer's environment using a digital screen, such as a smartphone or virtual reality headset, for example, the smartphone game, Pokémon Go; commonly abbreviated AR

Imaging An umbrella term referring to the creation of images typically through multimedia and digital means

Metaverse A growing network of virtual reality spaces, where users can interact virtually, typically as virtual avatars in real time via VR headsets or through a web browser

New media A broad term encompassing forms of audiovisual media that are technology oriented and rely on computers and the internet for redistribution, such as video, animation, and computer files

Postphotographic Describing an artistic theory that reevaluates humans' relationships with images and the methods used to produce them as technology changes interactions and interpretations of images

The sublime Generally, the quality of greatness, often found in the beauty and power of nature, its grandeur beyond comprehension; often associated with an existential experience where individuals consider their own mortality and purpose

Web 3.0 The third generation of the internet, in part associated with advancements in VR and digital 3D integrations

Virtual reality Computer graphics-generated images, 3-D objects, and environments that can interact through virtual reality-viewing equipment, such as headsets and haptics

CHAPTER 13

Expressive STEM Storymaking

Art, Literacy, and Creative Computing

Sean Justice and Lori Czop Assaf

Abstract

This chapter features intersections of art, literacy, and creative computing. As a component of STEAM, creative computing augments story creation, or storymaking (Buganza et al., 2023; Compton & Thompson, 2018), prompting learners to explore expressive meaning making as collective interactions with texts. To signify a way of teaching that supports such learning activities, we propose expressive STEM as a design principle, illustrated here with examples from an elementary school and a preservice art education program in Texas, USA. Principles of expressive STEM storymaking drawn from these examples and from our teaching and research are offered in the chapter's conclusion.

Keywords

art education – computational literacies – computational thinking – creative computing – storymaking

1 Introduction

This chapter features intersections of art, literacy, and creative computing. Working with new media and computational tools for creative purposes has become increasingly important to contemporary art education practices (Mohanty, 2022; Peppler & Wohlwend, 2018). In fact, "art educators are in the unique position to ask probing and provocative questions about digital media" (Sweeny, 2022, p. 293); but those in the field at large perceive a need for more substantive engagement with creative computing in universities and schools (Knochel & Patton, 2015; Leonard, 2021; Patton et al., 2020). This misalignment represents an opportunity to expand the relevance of creative computing in STEM–STEAM learning across multiple domains, including the

 | DOI:10.1163/9789004714748_014

arts, particularly since computational thinking (CT) has been situated as a core priority in K–12 education (Kafai & Proctor, 2022).

In the field of computer science education, CT has been framed as a way of thinking focused on representing problems as step-by-step procedures, so that those procedures can be programmed and carried out by a computer (Grover, 2022; Wing, 2017). Comprising skills derived from mathematics, engineering, and design that are "deeply symbiotic" (Grover & Pea, 2018, p. 33) with STEM disciplines, CT is a "foundational competency for being an informed citizen" (p. 34). Our focus here is not on isolated computing skills or competencies but on expanding literacy learning activities, particularly story creation, or storymaking (Buganza et al., 2023; Compton & Thompson, 2018), to include creative computing—a goal that aligns broadly with the reorientation of CT in computer science education as a component of socially transformative computational literacies (Kafai & Proctor, 2022).

Teachers of young children from 5 to 7 years of age take their responsibility for teaching literacy seriously, frequently telling us they will work with any tool that increases engagement and student success, yet they view adding computing to their repertoire with reservation. For instance, teachers and preservice students often begin workshops and courses declaring their distrust of computers. "Computers hate me!" is not an uncommon refrain; nevertheless, teachers are generally enthusiastic about the potential for STEAM activities to increase student engagement (Graham, 2021).

To discuss the expressive potential of storymaking + creative computing, we will use the term "expressive" STEM. To illustrate expressive STEM as storymaking, we describe three learning activities drawn from our work as teacher educators and CT researchers. The first is from a kindergarten classroom with 5- to 6-year-olds, where students responded to a read-aloud about camouflage by making an animated hide-and-seek word game in ScratchJr. The other examples come from a preservice digital methods course, where preservice students engaged in storymaking via computer programming and artmaking.

2 STEM–STEAM + Literacies

From preservice art education courses (Justice, 2020) to computer science education research (Justice & Assaf, 2020) to community-centered education (Families Learning Together, n.d.), we have noticed that learners gravitate to activities emphasizing interpersonal relationships, for example peer–peer, parent–child, teacher–student, even if those activities involve unfamiliar tools and materials. Activities aligned with STEAM learning include light-up Mothers

Day cards with LEDs and coin cell batteries, machines that make marks or noise to express the personality of a fictional character, and interactive oral histories programmed on a computer. As explored elsewhere (Justice, 2016, 2020, 2024), learner motivation for such activities appears to be driven not simply by a desire for STEM skills, for example, to improve math or science or computer programming abilities, but also by a desire for strengthening friend and family relationships. This observation is supported by a scoping review of STEM–STEAM and makerspace learning by Johnston et al. (2022), who found a noticeable increase in recent studies arguing for STEM learning centered on literacy, play, and family.

Lately, though, we have become skeptical of identifying with either STEM *or* STEAM because these acronyms appear too simplistic. At root, we question integration as a turnkey principle (e.g., simply inserting art between engineering and math) because it minimizes the way learning matters for learners. This realization reprises something we noticed a long time ago but perhaps did not fully understand at the time. In 2014 during a study of K–12 digital making and learning (Justice, 2016), a director of technology complained that focusing on integration was misguided if it centered on tools rather than on teaching. He said, "If [the teaching] is strong with the FabLab, that's great; but if it's strong without the FabLab, that's great too" (as cited in Justice, 2016, p. 89).

This shift toward *how* learning with tools matters, without fixating on the tools themselves, challenges teaching methods. When we introduce creative coding as a way for students to respond to stories (e.g., Justice, 2019), teachers and preservice students sometimes protest that meaningful engagement cannot occur without deep scaffolding, that is, without teaching the basics of computer programming with step-by-step tutorials. After implementing such activities in classrooms, without becoming experts themselves, however, those same teachers and preservice students are astonished when children become completely absorbed with drawing, coding, and narrating stories on a computer or tablet.

3 Computing as Expressive

Framing computing as expressive follows the positioning of CT in computer science (CS) education as transdisciplinary, a domain without its own content (Grover & Pea, 2018). Grover (2022) argued that CT skills-learning is stronger when embedded in diverse content domains, both in CS and non-CS classrooms, and when teaching is multimodal, that is, taught in a variety of ways with a variety of tools and materials. For Grover, multimodality positions CT "*so that skills developed are conceptual and creative*" (p. 20, emphasis in the original); furthermore, "learning CT in different ways for different purposes, including creative expression, helps diverse learners engage meaningfully" (p. 34). Likewise, Kafai and Proctor (2022) recognized that situating CT across domains frames "computing as a vehicle for personal expression and connecting with others alongside and intersecting a plurality of other literacy practices" (p. 148).

This emphasis on computing as expressive is not unprecedented. Even though educators often associate computers with cold, depersonalized machinery, Denning and Tedre (2019) stated that computation depends on human ways of thinking that have emerged over thousands of years, anchoring CT in the bedrock of human expression. Similarly, Resnick (2006), the MIT Media Lab computer scientist who created the Scratch programming language, has argued "computers will not live up to their potential until we start to think of them less like televisions and more like paintbrushes" (p. 192); and Grover and Pea (2018) have claimed that "from music, math, social studies, history, language arts, and throughout the sciences and engineering, curricular ideas can come alive with CT" (p. 32).

4 Expression as Response

Art and literacy educators understand expressive response as a powerful learning principle relying on coordinated, multimodal partnerships. Art educator Hafeli (2015) proposed that artists begin by noticing the way various tools and materials express mood and concept differently and then learn to work with those differences to elicit ideas and emotions. For example, marks made with hard graphite can feel sketchy or tentative, unsure of themselves, while marks made with soft charcoal might evoke authority, boldness, or anger. Noticing these potentials, or affordances, casts artists as responding *with* tools and materials—in partnership—rather than coldly *using* them. With a similar emphasis on diverse material affordances, Pacini-Ketchabaw et al. (2017) described expressive response as an *encounter* of the world, implying serendipity, or a sense of not entirely preplanned, an experience of being in the world that counters the predictable intentionality implied by conventional notions of artistic control. Instead, Hafeli (2015) and Pacini-Ketchabaw et al. (2017) emphasized artistic meaning making as a coordination of diverse voices, where the artist's task is more about feeling the pulse of experience and calling forth an expressive response *with* those voices rather than telling them what to do or say.

Similarly, literacy teachers Kuby and Rucker (2016) identified writing as a multimodal response comprising diverse practices, not limited to reading and writing. In fact, in the field of literacy education, writing has long been identified as plural, that is, as literacies, or multiliteracies (New London Group, 1996). For Kuby and Rucker, a multiliteracies approach to the teaching of writing positions reading as engaging with the "world through traditional means such as books, but also images, digital encounters, performance, and art" (p. 12). In their view multimodal writing with murals, videos, and puppets sparked "literacy desirings" (p. xv), enhancing student engagement and learning.

To explain how multimodality increased their 7- to 8-year-old students' interest in writing, Kuby and Rucker (2016) pointed to the availability or scarcity of tools and materials in the classroom. For example, Rucker worried she might have discouraged literacy desiring because she "had never made [materials] freely available" (p. 62) for students' responses. Encouraging puppet-making in a writing workshop, on the other hand, permitted a serendipitous frog puppet to catalyze students' desire for "bridging literacies at home with literacies at school" (p. 12). Here, Kuby and Rucker explained that crossing thresholds between school and not-school created a transdisciplinary plurality that augmented individual and community literacies, a desire for sharing in-school learning with family and friends outside school, which reciprocally enhanced engagement back in the classroom.

5 Storymaking

Across domains and learning spaces, with early childhood students 5 to 8 years of age and up to adult learners in professional organizations, storymaking is defined as a playful, exploratory mode of story creation that shapes and sustains collective identity, learning, reflection, and belonging (Buganza et al, 2023; Bunda et al., 2019; Compton & Thompson, 2018; Smeed, 2012). From the perspective of adult learning in organizational development, Buganza et al. (2023) defined storymaking as a bidirectional "collective process of sensemaking" (p. 11), a "discursive tool" for creating "new meanings and shaping intrinsic identities" (p. 12, Table 2.1). By contrast, Buganza et al. described story*telling* as a monodirectional, top-down approach, useful for getting information across.

For Bunda et al. (2019), a research and artmaking collective of Australian women from diverse backgrounds, storymaking fosters collaborative responses to absent or incomplete individual and community histories. In their artmaking the collective brought together diverse materials to reclaim a sense of belonging: "Black and White Australian women provoked resonant and entangled understandings of belonging and displacement through performative making" (p. 158), producing multimodal archives comprising photographs, sculptures, baskets, poems, and stories.

For early childhood learning specialists Compton and Thompson (2018), storymaking comprises expressive responses to "multimodal texts (language, art, acting, connecting, drawing, collage)" (p. 13). When they are storymaking, children work as artists, writers, engineers, and performers to express ideas that matter to them. By sharing storymaking, children expand their own and their community's literacies as interactive, intersecting histories connected to friends, peers, classrooms, and families. Thus, Compton and Thompson have argued, "storymaking creates an inclusive community [where] students come to know themselves and one another" (p. 16). With a nod to Reggio Emilia's child-centered inquiry methodology, Compton and Thompson positioned storymaking as interactions between text and materials that "can include any of the Hundred Languages referred to by Loris Malaguzzi, such as drawing, sewing, painting, sculpting, weaving, dramatic play, music, and dance" (p. 15).

From a historical perspective, Compton and Thompson's (2018) emphasis on the interactional multimodality of storymaking reprises Louise Rosenblatt's (1933/1995) reader response theory, positing meaning making as an interaction between reader and text, not as an isolated property of either component. This notion of interactivity—where meaning making is relational and responsive to the interdependency of reader and text—is consistent with John Dewey's (1934/1980) theory of art as experience, published at roughly the same time. In fact, relational theories of meaning making rooted in experience and response from the early 20th century—like Rosenblatt's and Dewey's—continue to inform approaches to art and literacy in schools (e.g., Hafeli, 2015; Kuby & Rucker, 2016).

6 Storymaking as Expressive STEM

In our university courses and professional learning workshops with teachers, expressive STEM storymaking invites participants to explore story creation with computational tools and materials, sometimes including traditional craft tools and art materials. Learners respond to stories of their choice by focusing on one or two major story elements, such as character, setting, or plot, often centering on early elementary storybooks. The goal is to stage a story or part of a story by inventing an original narrative, by transforming an existing narrative, or by responding to domain prompts from language arts, science, math, or social studies. Elementary and preservice teachers without much experience with creative computing work with computer programming platforms designed for novices, such as Scratch[1] or Scratch Jr.[2] In art education settings the emphasis is on how design choices convey mood and concept with new media principles (e.g., visual, aural, and time-based components) combined with conventional elements and principles of art (e.g., color, shape, balance).

The examples below describe expressive STEM in practice. Example 1 is from a kindergarten teacher who participated in a CT research project (Justice & Assaf, 2020); Examples 2 and 3 are from preservice art education students who participated in digital learning courses.

6.1 *Example 1: Storymaking in Kindergarten*

As part of a computer science research project, looking at classroom teachers' implementation of CT (Justice & Assaf, 2020), we observed a kindergarten teacher invite her 5- to 6-year-old students to code a computational hide-and-seek animation with words that are difficult for young readers to decode, commonly known as sight words, for example, "we," "the," "good" (Figure 13.1). After reading *Ruth Heller's How to Hide a Butterfly & Other Insects* (Heller, 1992), Bridget (a pseudonym) asked the children to respond with ScratchJr., a block-based computer programming platform designed for early readers. The book emphasized the camouflage strategies insects use in their habitats, so Bridget's guided practice focused on the word "camouflage" by asking the children to name things in their lives—that is, in their own habitats—that were difficult to see. The goal was for children to connect "habitat" and "camouflage" with

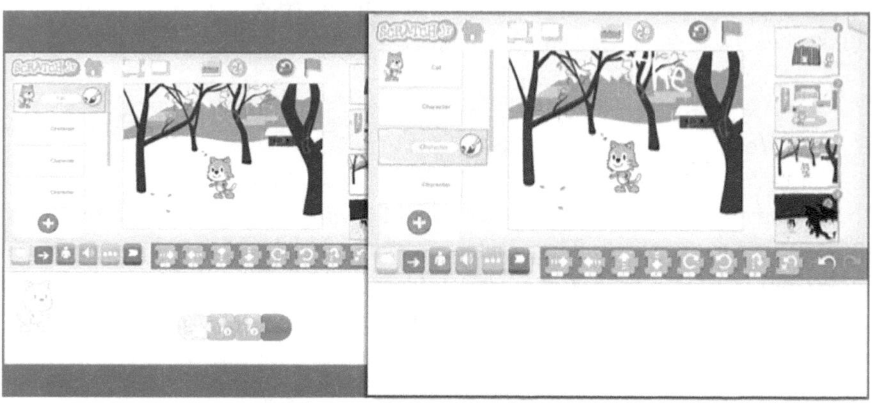

FIGURE 13.1 Camouflaged sight words in ScratchJr. The right screen shows a sight word hiding in the trees. The left screen shows the student's code: when the cat is touched, ScratchJr. will play audio made by the student of herself speaking the sight words she has hidden

things in their world and then to expand their notion of things to include sight words.

To motivate the children, Bridget talked about hiding sight words in ScratchJr. by adjusting color or scale until the words blended into the backgrounds. She also encouraged students to explore other camouflage strategies. Afterward, sitting cross-legged on the carpet, shoulder to shoulder with their partners, the children shared their strategies and identified the sight words they had chosen to camouflage.

On prior visits to her classroom, we had noticed Bridget prompting students to respond to texts with various tools and materials by drawing pictures or making collages with paper, markers, glue, tape, and miscellaneous craft materials. Here, Bridget had innovated her approach by including ScratchJr; that is, despite feeling like a novice with CT—having only recently encountered computer programming as part of our research project—Bridget felt confident enough in what she already knew about teaching and learning to trust the children to find their way with this unfamiliar tool.

And from what we could see, her innovation paid off. Every child appeared engaged and on task, playing with sight words, hiding and then revealing them against multicolored backgrounds. As we listened to their conversations, we noticed children programming animations to represent their thinking about sight words by adapting what they already knew about color, shape, texture, and scale; by comparing their knowledge of sight words with their classmates' knowledge; and by troubleshooting their projects collaboratively with their partners.

A few weeks later Bridget reflected on the lesson during a Zoom interview. When asked what she thought about how the lesson had gone, she said it had been "a home run out of the park, [because] the kindergartners [now] use the word 'camouflage' every day." (all quotations in this paragraph and the next are from Bridget, personal communication, April 7, 2022). When asked how she felt about her teaching overall, she said she had struggled to step back from instructional methods that rely on telling students what to do, that is, rather than inviting them to explore and play. She said, "It's very difficult. I've been taught to be a teacher in a certain way, and that's direct teach. That's what's expected." The challenge, she explained, was that telling students what to do might "ruin their learning moment." And she said, "I've been trying so hard to not just swoop in and be like, 'Look, let me show you how to do it—you click this, you go there, and you do this. Now, show me what you can do.' I don't want to do that. I want them to learn this on their own."

But then she added that she was committed to inquiry learning despite her misgivings because she and her students were learning together. She explained, "We already know that whenever you find something on your own and you learn how to do it, it's intrinsically woven into the fabric of your being."

6.2 *Example 2: Storymaking on the Screen*

In a digital learning course for preservice teachers, Jill Picou, a former art education student who teaches 11- to 13-year-old students in middle school, responded to a storymaking prompt with an adventure story about knights in training (Figure 13.2).

Knights of the Woods (Picou, 2021) presents an original story about a sister and her brother playing in the forest near their home. The story opens with them walking across a forest meadow. As Oliver skips along, he suggests playing knights. Emma takes charge—“Follow me!”—and imagines a training program for knighthood. “Knights are brave,” she says, balancing atop a large rock. Later, she and her brother fight a dragon to protect the castle. The story is interrupted when the Queen (Emma and Oliver’s mother) calls them home for lunch, at which point the castle turns back into a pile of rocks in the forest and the dragon transforms into Blaze, their golden retriever. The story ends with Emma, Oliver, and Blaze traipsing back across the meadow.

Knights of the Woods combines hand-drawn characters and original dialog with computer code that animates characters and backdrops. To amplify audience participation, Picou also programmed interactive elements into the story—for example, tapping keys on the keyboard propels the narrative. On the project homepage, Picou wrote instructions for manipulating the interactive elements as well as a list of credits. Inside the project, she explained her

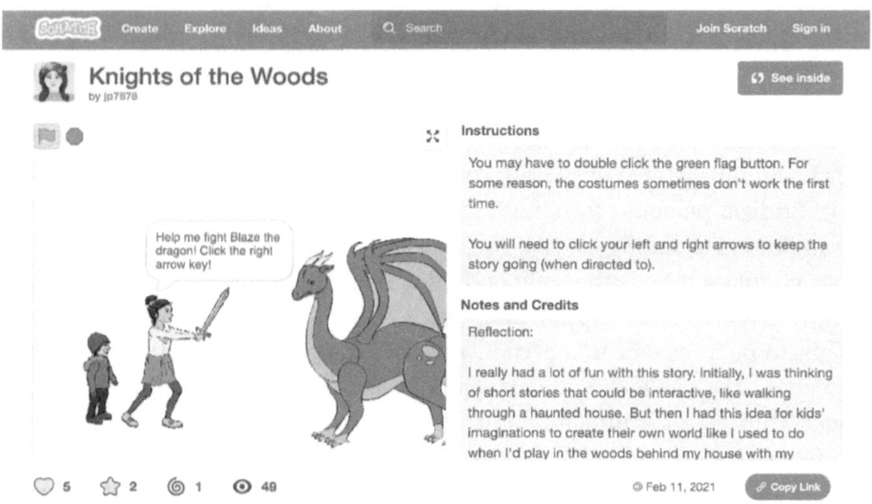

FIGURE 13.2 Interactive, screen-based storymaking. *Knights of the Woods* (Picou, 2021) combined hand-drawn characters with Scratch’s stock characters and dialog, music, and creative coding to advance the narrative. To amplify reader interaction, Picou programmed keystrokes that animate the characters when tapped by a reader

computational design decisions, so readers can follow her work or borrow her code.

As evident here, Picou deployed multiple storymaking strategies, including drawing, collaging, coding, sound recording, and written reflection. This robust multimodality deepens reader interaction beyond the narrative itself. For example, on the project homepage Picou explained she had considered creating a story based on a fantasy setting like a haunted house but changed her mind after reflecting on the power of children's "imaginations to create their own world, like I used to do when I'd play in the woods behind my house with my friends" (Picou, 2021, para. 1). Inside the project Picou discussed her learning trajectory: "I learned a lot in the process of creating this story including how to create a loop of movement (changing costumes) until the sprite reached a certain x point or the backdrop changed" (para. 5). She also reflected on art, learning, and computation: "We can draw all we want, but if you can't figure out a way to create movement, then it's not animation. You have to flex both of those brain muscles and not give up on it" (para. 6).

In 2023, while writing this chapter, we asked Picou whether making *Knights of the Woods* still resonated for her now that she was an art teacher. She said it had been one of her first creative computing experiences, and that before the digital learning course, she had never worked with computer code. Perhaps because of that, she said, the storymaking prompt was valuable because "instead of just learning the specific sequence of codes, I had a goal. I was intrinsically invested in learning how to code in Scratch so I could create the story I had in mind" (Picou, personal communication, January 9, 2023).

Regarding her evolving teaching practice, Picou said story creation was woven into her art lesson plans because it enabled students to express their experiences. She added that art and literacy connections were especially important for emergent bilingual students because they strengthened written reflection, language use, and vocabulary. Beyond that, she said, positive responses to challenges like expressive storymaking helped improve students' self-efficacy. This was particularly important because, she said, her "current students seem to give up on their goals very easily when they run into the first roadblock, ... [so] the whole process of troubleshooting (in coding or in any other project) is hugely applicable to [them]" (Picou, personal email communication, January 9, 2023).

6.3 *Example 3: Storymaking as Transdigital*

If You Are Given a Cookie ... by Lyla Guidry (2022), a preservice art education student, is an interactive digital + physical, or transdigital (Fuglestadt, 2022), response to *If You Give a Mouse a Cookie*, written by Laura Numeroff (2015) and

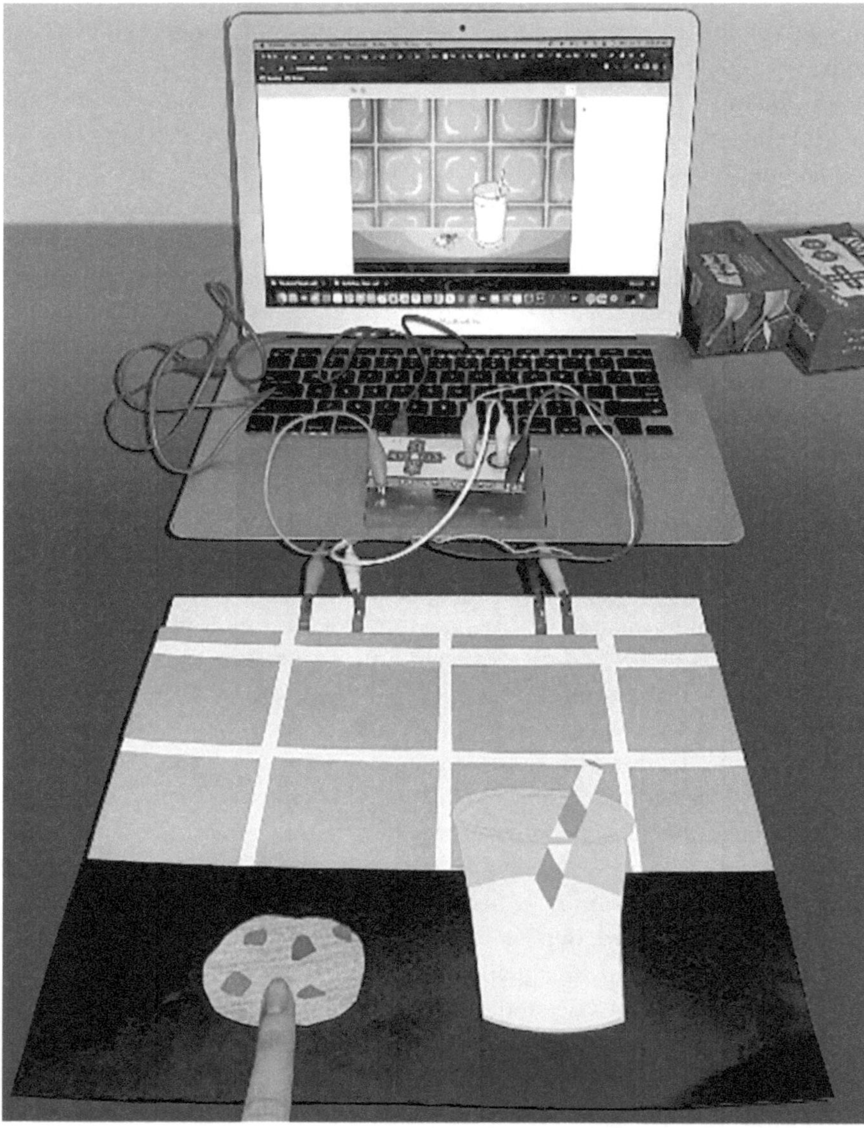

FIGURE 13.3 Transdigital storymaking. *If You Are Given a Cookie …* by Lyla Guidry (2022) is constructed from a paper collage connected to a Scratch animation with a Makey Makey®. Each touch of the paper cookie takes another bite from the screen cookie, producing a munching sound

illustrated by Felicia Bond. Guidry's project included a screen-based animation programmed on Scratch connected to a construction paper collage with embedded electronic switches made with copper tape and a separate circuit board, known as a Makey Makey® (https://makeymakey.com/; Figure 13.3).

The screen-based component can be found on Scratch by searching for its title on the Scratch homepage, but the transdigital components (the collage, the embedded switch, and the Makey Makey®) will not be visible there.

To explore storymaking, Guidry responded to the plot circularity of *If You Give a Mouse a Cookie*: Cookie leads to milk, mirror, broom, nap, story, picture, milk, and back to cookie. With the circuit-collage assembled and attached to the computer, a reader touches the paper cookie (on the collage) to take a bite out of the cookie (on the screen), producing a munching sound. Each touch of the paper cookie takes another bite from the screen cookie, and we hear another munch. After several munches nothing remains of the screen cookie except crumbs, but when the paper cookie is touched once more, the screen cookie reappears, and the touch–munch cycle can begin again. A similar touch–slurp cycle is activated when the glass of milk is touched.

In this transdigital activity storymakers coordinate diverse tools and materials (i.e., screen-based animation, paper collage, electric circuitry) to amplify reader–text interaction. In fact, the story itself invites reader interaction because physical touch advances the narrative. In addition, making the story insists on *writer* interaction because aligning these diverse components requires multiple rounds of designing, building, and troubleshooting.

As with Picou we contacted Guidry while writing this chapter to ask what she remembered about the storymaking project. She said that getting started had made her feel confused and a little disheartened. But then, while watching a classmate fold construction paper to interact with a book about butterflies, transforming paper into butterfly wings, Guidry had a breakthrough. She said that sharing that tiny bit of her classmate's experience produced a spark that lit up her own storymaking desire, and recognizing her classmate's deeper understanding prompted her to dig deeper for herself. Nearly a year after Guidry had taken the digital learning course, the power of that moment still resonated for her.

Specifically, regarding her own storymaking experience, Guidry remembered deciding to respond to a single element from the book and let the rest flow from there. That element turned out to be narrative circularity as expressed by the plot of *If You Give a Mouse a Cookie*. Guidry then created her story response centered on loops, a powerful idea from computer programming (Bers, 2021) that the class had explored in prior computing activities. Subsequently, she said, creating loops in Scratch and connecting the code to the collage via copper tape and the Makey Makey® felt childlike—like a playful exploration of eating and drinking again and again.

We also asked Guidry whether expressive STEM storymaking felt relevant to becoming an art teacher. She said storymaking connected domains that appeared independent by aligning diverse toolsets. For Guidry the activity helped her understand how a teacher might introduce unfamiliar materials to students while focusing on expressive learning and meaning making. With a reference to traditional art making tools, Guidry said, "Most materials I work with are normal to me, [so the storymaking project] humbled me a little bit since I hadn't been in that position in years" (L. Guidry, personal communication, January 9, 2023).

7 Principles of Expressive STEM Storymaking

After reflecting on expressive STEM storymaking with young children in schools, families at public libraries, preservice art education students at the university, and teachers in professional learning workshops, we have noticed that balancing the contrasting and sometimes conflicting affordances of new and traditional tools and materials can be challenging. For example, in a transdigital activity, preservice art students might be comfortable expressing themselves with picture-making materials but frustrated by copper tape, circuit switches, Makey Makey®s, and Scratch. We see the same thing but in reverse with computer science students who hesitate to engage with drawing or collaging. The situation is similar with most other learners, too, the first time they encounter Scratch or ScratchJr. or any number of other expressive STEM materials. Nevertheless, as illustrated in the examples above, learners dig in, become deeply engaged, play, and explore purposefully, iterate and troubleshoot, adapt themselves to failure, and above all, share the storymaking process with their community of learners.

The following principles of expressive STEM storymaking have been gleaned from more than 10 years designing and implementing creative computing learning activities. Some are evident in the examples described above, but others have not been explicitly discussed; for example, we did not describe the classroom setup and only briefly mentioned the importance of *noticing* rather than *assessing* as a way to spark storymaking desires. Thus, the list is not a catalog of best practices; instead, along with resources listed in the Appendix, the list might help as a framework for starting with expressive STEM storymaking even for teachers with little experience with creative computing.

8 Ten Guiding Principles for Expressive STEM

1. Invite responses to multimodal text(s) as an experience of meaning making rather than as an exercise of tool use.

2. Encourage exploration of the way tools and materials express differently, for example, hard graphite vs. soft charcoal; fast, skittery motion vs. slow, smooth motion; looped percussion vs. continuous melody; interactive touch via the keyboard vs. a transdigital collage, also known as an *interface* (Justice 2019, 2020).
3. Facilitate collaborative interactions between learners and story elements or domain prompts by focusing on interpersonal relationships; for example, rather than focusing on simple appearances, ask, "How would it feel to be in this setting or with this character?"
4. Minimize instructional talk to allow time for exploration, iteration, and purposeful play with examples, tools, and materials.
5. Maximize multimodality of learning resources, for example, demonstrations plus videos plus handouts plus websites plus instructional manuals plus safety data sheets and more.
6. Design participant-led show-and-shares that emphasize noticing instead of assessment; for example, after sharing their work, learners ask each other, "What do you notice about my project?"—not, "What do you like?"
7. Arrange the classroom or studio with large tables instead of individual workstations to enhance learners' interactions with one another.
8. Share models and examples instead of step-by-step instructions.
9. Make sure tools and materials are visible, plentiful, and multimodal, for example, digital, computational, and traditional craft or fine art.
10. Intersperse reflective turn-and-share moments with uninterrupted making and building.

9 Acknowledgments

Thank you, Jill Picou and Lyla Guidry, former art education students at Texas State University, for allowing us to share your work here. The work reported in this chapter has been partially funded by the National Science Foundation, Grant # 2006595. Any opinions, findings, and conclusions or recommendations expressed here are those of the authors and do not necessarily reflect the views of the National Science Foundation.

Notes

1 https://scratch.mit.edu
2 https://www.scratchjr.org/

References

Bers, M. U. (2021). *Coding as a playground: Programming and computational thinking in the early childhood classroom* (2nd ed.). Routledge.

Buganza, T., Bellis, P., Magnanini, S., Press, J., Shani, A. B., Trabucchi, D., Verganti, R., & Zasa, F. P. (2023). *Storymaking and organizational transformation: How the co-creation of narratives engages people for innovation and transformation*. Routledge.

Bunda, T., Heckenberg, R., Snepvangers, K., Phillips, L. G., Lasczik, A., & Black, A. L. (2019). Storymaking belonging. *Art/Research International, 4*(1), 153–179. https://doi.org/10.18432/ari29429

Compton, M. K., & Thompson, R. C. (2018). *StoryMaking: The maker movement approach to literacy for early learners*. Redleaf Press.

Denning, P. J., & Tedre, M. (2019). *Computational thinking*. The MIT Press.

Dewey, J. (1980). *Art as experience*. Perigee Books. (Original work published 1934).

Families Learning Together (n.d.). https://fltsmtx.wp.txstate.edu/

Fuglestad, T. (2022). Make it, move it, flip it. *SchoolArts, 121*(7), 36–37.

Graham, M. A. (2021). The disciplinary borderlands of education: Art and STEAM education. *Journal for the Study of Education and Development, 44*(4), 769–800. https://doi.org/10.1080/02103702.2021.1926163

Grover, S. (2022). Computational thinking today. In A. Yadav & U. D. Berthelsen (Eds.), *Computational thinking in education: A pedagogical perspective* (pp. 18–40). Routledge. https://doi.org/10.4324/9781003102991

Grover, S., & Pea, R. (2018). Computational thinking: A competency whose time has come. In S. Sentance, E. Barendsen, & C. Schulte (Eds.), *Computer science education: Perspectives on teaching and learning in school* (pp. 19–38). Bloomsbury Academic. https://doi.org/10.5040/9781350057142.ch-003

Guidry, L. (2022, March 27). *If you are given a cookie ....* Scratch. https://scratch.mit.edu/projects/664121832/

Hafeli, M. (2015). *Exploring studio materials: Teaching creative art making to children.* Oxford University Press.

Heller, R. (1992). *Ruth Heller's how to hide a butterfly & other insects*. Grosset & Dunlap.

Johnston, K., Kervin, L., & Wyeth, P. (2022). STEM, STEAM and makerspaces in early childhood: A scoping review. *Sustainability, 14*(20), Article 13533. https://doi.org/10.3390/su142013533

Justice, S. (2016). *Learning to teach in the digital age: New materialities and maker paradigms in schools*. Peter Lang.

Justice, S. (2019). Interface: The transformative potential of computational making. *NAEA News, 61*(5), 20. https://doi.org/10.1080/01606395.2019.1657759

Justice, S. (2020). Designing the social interface: More than social, more than material. In A. D. Knochel, C. Liao, & R. M. Patton (Eds.), *Critical digital making in art education* (pp. 61–74). Peter Lang.

Justice, S. (2024). Engineering joy: Toying with process in expanded media arts. In T. Hunter-Doniger & N. Walkup (Eds.), *STEAM education: Transdisciplinarity of art in the curriculum* (pp. 118–129). Routledge.

Justice, S., & Assaf, L. (2020). *Exploring early childhood teachers' abilities to identify computational thinking precursors to strengthen computer science in classrooms* [Abstract]. National Science Foundation. https://www.nsf.gov/awardsearch/showAward?AWD_ID=2006595

Kafai, Y. B., & Proctor, C. (2022). A revaluation of computational thinking in K–12 education: Moving toward computational literacies. *Educational Researcher, 51*(2), 146–151. https://doi.org/10.3102/0013189x211057904

Knochel, A. D., & Patton, R. M. (2015). *If* art education *then* critical digital making: Computational thinking and creative code. *Studies in Art Education, 57*(1), 21–38. https://doi.org/10.1080/00393541.2015.11666280

Kuby, C. R., & Rucker, T. G. (2016). *Go be a writer!: Expanding the curricular boundaries of literacy learning with children*. Teachers College Press.

Leonard, N. (2021). Emerging artificial intelligence, art and pedagogy: Exploring discussions of creative algorithms and machines for art education. *Digital Culture & Education, 13*(1), 20–41.

Mohanty, K. (2022). Media arts [Co-editor's letter]. *SchoolArts, 121*(7), 8. https://lsc-pagepro.mydigitalpublication.com/publication/?m=60985&i=736878&p=10&ver=html5

New London Group. (1996). A pedagogy of multiliteracies: Designing social futures. *Harvard Educational Review, 66*(1), 60–92. http://dx.doi.org/10.17763/haer.66.1.17370n67v22j160u

Numeroff, L., & Bond, F. (2015). *If you give a mouse a cookie*. Harper Collins.

Pacini-Ketchabaw, V., Kind, S., & Kocher, L. L. M. (2017). *Encounters with materials in early childhood education*. Routledge.

Patton, R., Sweeny, R. W., Shin, R., & Lu, L. (2020). Teaching digital game design with preservice art educators. *Studies in Art Education, 61*(2), 155–170. https://doi.org/10.1080/00393541.2020.1738165

Peppler, K., & Wohlwend, K. (2018). Theorizing the nexus of STEAM practice. *Arts Education Policy Review, 119*(2), 88–99. https://doi.org/10.1080/10632913.2017.1316331

Picou, J. (2021, February 11). *Knights of the woods*. Scratch. https://scratch.mit.edu/projects/485857870/

Resnick, M. (2006). Computer as paintbrush: Technology, play, and the creative society. In D. Singer, R. Golikoff, & K. Hirsh-Pasek (Eds.), *Play = learning: How play motivates and enhances children's cognitive and social-emotional growth* (pp. 192–206). Oxford University Press.

Rosenblatt, L. M. (1995). *Literature as exploration*. Modern Language Association. (Original work published 1933)

Smeed, J. (2012). The grumpy dragon and the angry dragon: From storytelling to storymaking. *Storytelling, Self, Society*, *8*(1), 1–16. https://doi.org/10.1080/15505340.2012.635092

Sweeny, R. (2022). Review of the book *Critical digital making in art education*, by A. D. Knochel, C. Liao, & R. M. Patton, Eds. *International Journal of Education Through Art*, *18*(2), 289–294. https://doi.org/10.1386/eta_00101_5

Wing, J. M. (2017). Computational thinking's influence on research and education for all. *Italian Journal of Educational Technology*, *25*(2), 7–14. https://doi.org/10.17471/2499-4324/922

Appendix

Resources for getting started with expressive STEM storymaking activities.

Creative Computing

- Scratch. https://scratch.mit.edu/
  Create stories, games, and animations. Share with others around the world.
- ScratchJr. https://www.scratchjr.org/
  Coding for young children.
- Creative Computing Curriculum. http://scratched.gse.harvard.edu/guide/
  Ideas, strategies, and activities for an introductory creative computing experience using Scratch.
- Scratch Foundation. https://www.scratchfoundation.org/
  Supporting creative coding for everyone.

Transdigital Making

- Makey Makey®. https://makeymakey.com/
  Invention kit for the 21st century. Connect the world to your computer!
- micro:bit. https://microbit.org/
  Get creative, get connected, get coding! The pocket-sized computer transforming the world.
- the Interface. http://seanjustice.com/interface/
  This website contains instructions for a transdigital computational making activity.

CHAPTER 14

Considering STEAM Aesthetics through an Arts-Driven STEAM Pedagogy

Alice Lai

Abstract

This chapter features an arts-driven STEAM (ADSTEAM) pedagogical model I devised aligning with the historical and Bauhaus views of the symbiotic relationships among STEAM subjects. Using my undergraduate theory of art class as a case study, I detail the application of the ADSTEAM model, including the pedagogical objectives and strategies. After elucidating students' probing questions concerning aesthetic theories using the framework of STEAM, I delineate a studio activity prompting students to apply the Fibonacci mathematical patterns to create visual work. Three students' projects and artist statements demonstrating the transdisciplinary thinking underlying visual compositions, creativity, and aesthetic theories affirm the generous potential of ADSTEAM learning.

Keywords

ADSTEAM – Bauhaus – aesthetic theory – Fibonacci sequence

1 Introduction

STEAM education is considered a progressive and holistic educational paradigm for learners in the 21st century to gain interrelated content knowledge and academic skills across the disciplines of science, technology, engineering, arts and humanities, and mathematics (Burnard & Colucci-Gray, 2020; Hunter-Doniger & Walkup, 2024). In the field of art education in the United States, case studies of STEAM pedagogy are often derived from K–12 art classes and graduate-level classes designed for preservice and in-service K–12 teachers. In comparison, research on STEAM pedagogy in undergraduate arts and humanities classes in which students of art education develop foundational knowledge and skills in the arts is limited (Lai, 2024).

 | 10.1163/9789004714748_015

At my university, studio art, art appreciation, and art theory classes offer the fundamentals for undergraduate students aspiring to become art teachers; however, as a former associate department chair reviewing and overseeing the syllabi and classes in the department for 20 years, I have noticed few STEAM activities therein. Educational researchers posit two adverse conditions leading to these shortfalls: first, reservations that arts and humanities professors have about their ability to blend science, technology, engineering, or mathematics into their curriculum and second, the silo phenomenon, both creating barriers for the professors to traverse discipline-based confinement and obligations to embrace the STEM or STEAM paradigm (Cohen & Lloyd 2014; Herro et al., 2019). As a result of these challenges, not only more advocacy for but also empirical studies and longitudinal research on STEAM teaching in foundational visual arts curricula are needed to motivate and assist arts and humanities professors to adopt the STEAM paradigm in their teaching.

When designing a STEAM curriculum, conceptually, I adhere to the principle of placing art at the forefront and heart of the STEAM curriculum, rejecting the treatment of art as a compensatory, peripheral, or motivational subject (Pirrie, 2019). I also adopt STEAM education as a transdisciplinary learning paradigm neither aimed at fostering single discipline-based learning nor demanding mastery of knowledge in all involved disciplines. Instead, STEAM learning foments integrative knowledge construction and manifestation, blending necessary knowledge and skills from involved disciplines and resulting in innovation transcending canonical scholarship and making of art (Areljung, 2023; Liao, 2016; Marshall, 2014).

Pragmatically, I consider two questions: How might STEAM knowledge be featured in arts and humanities classes to broaden students' appreciation of art beyond the conventional art discipline-based content and mode? How might STEAM principles be infused into studio art activities in a way that augments students' creative thinking, quality of artworks, and potential in artistic innovation? Through iterative experimentation, my curriculum now motivates students to study relevant knowledge and skills of science, technology, engineering, or mathematics, contributing to refreshed and complex comprehension of arts and humanities. Ultimately, I strive to cultivate in students a transdisciplinary ability to articulate the intersections of STEAM subjects, to analyze art, and to create original works of art.

In this chapter I address the questions noted above by introducing the arts-driven STEAM (ADSTEAM) pedagogical model I devised. After surveying historical and Bauhaus views on the symbiotic relationships among STEAM subjects, I identify ADSTEAM pedagogical objectives and strategies. Taking my undergraduate theory of art class as a case study, I detail the application of the ADSTEAM model. First, by elucidating students' probing questions concerning aesthetic theories using the framework of STEAM, I next delineate a studio activity impelling students to create visual works guided by the Fibonacci mathematical patterns. Three students' projects and artist statements demonstrating transdisciplinary thinking of visual compositions, creativity, and aesthetic theories allow a glimpse into the considerable potential of ADSTEAM learning.

2 The Intersections of STEAM Subjects: Historical and Bauhaus Views

Archetypal thinking about the arts and humanities in various ages can attest to the synergistic relationships among STEAM subjects. From Thomas Aquinas (ca. 1225–1274), Leonardo Da Vinci (1452–1519), Immanuel Kant (1724–1804), and Clive Bell (1881–1964) to Bauhaus founder Walter Gropius (1883–1969), philosophers and artists have trespassed the boundaries of the arts, philosophy, religious studies, STEM, and business throughout the history of Western art to advance their theorization and creation of art (Bullot et al., 2017; Freeland, 2001; Siebenbrodt & Schöbe, 2009). Their work confirms that STEM knowledge, such as geometry, perspectives, ratio, the psychology of perception and mind, and life science, intersect with the theories and practices of art. For example, Aquinas' concept of beauty, involving proportion, harmony, and light, blended mathematical, aesthetic, and theological thoughts in his time. Leonardo, revered as a "Renaissance man," filled his notebooks with drawings, mathematical works, architectural renderings, diagrams of mechanical devices, and human anatomical illustrations—the STEAM-infused humanistic musings of his time.

Kant's aesthetic theory bridges cognitive science, linguistics, and philosophy to elucidate the intricate networking among humans' mental faculties when encountering a work of art. Kant and Bell both postulated that aesthetic pleasure arises when our mental faculties are provoked by the right arrangement of formal properties of art (e.g., lines, colors, and masses). Their work begat one of the most influential theories in art—formalism—which emphasizes the expressiveness of form, established and manifested through the interplay of the formal properties of art and the principles of design, including balance, harmony, unity, movement, proportion, and pattern. Influenced by the psychologies of visual perception along with its terminologies used in the principles of design, formalism features the concepts, language, and methods employed in STEAM.

In the 20th century, the Bauhaus school offered a prototype for integrating art and STEM into the social theory of art and art curriculum. Toward the end of World War I between 1919 and 1933, the architect Walter Gropius along with a group of esteemed artist–faculty members inaugurated a Germany school of architecture, applied arts, and design called the Bauhaus, meaning "house of building." Rejecting the confinement of art within the academic canon and context and breaking down the divide between art and craft, or fine and applied art, the mission of the Bauhaus was to cultivate forward-thinking artists who would embrace contemporary machines, technology, and scientific methods for creating modern-looking, rationally beautiful, and sturdy art enriching daily and social life. Gropius revolutionized modern aesthetics, bringing together scientific processes, technological experiment, engineering rigor, mathematical principles, and a pragmatic and entrepreneurial mindset to stimulate the creation of fine and applied artistic objects. Subsequently, the Bauhaus educational and aesthetic paradigm paved the way for the advancement of color theory and a wide range of modernist art, including abstract painting, textile design, electric kinetic art, graphic design, futuristic furniture design, and modern architecture. Although the school was short-lived because of its radical educational philosophy and design aesthetics that rattled the social and artistic dogma favored by the Nazi regime, the Bauhaus continues to shape the concept, purpose, and forms of contemporary art, craft, design, and architecture well into the 21st century (Bayer et al., 1938/2017; Siebenbrodt & Schöbe, 2009).

The inside flap of the *Bauhaus, 1919–1928* exhibition catalog included the following: "WHAT is the Bauhaus? The Bauhaus is an answer to the question: How can [artists] be trained to take [their] place in the machine age[?]" (Bayer et al., 1938/2017). Responding to this inquest, the faculty team at the Bauhaus envisioned art as a melding between rational thinking and the beautiful object and "that the study of rational design in terms of technics and materials should be only the first step in the development of a new and modern sense of beauty" (Barr, 1938/2017, p. 8).

The intersections of art and STEM indeed formed the foundation of the Bauhaus art curriculum. For example, Wassily Kandinsky (1866–1944) in his Analytical Drawing class, asked the students to draw a still-life not only by exercising keen observation of the objects but also by dissecting the forms of the objects using a grid. Employing scientific and engineering techniques, Kandinsky's pedagogy challenged students to reconceptualize and revisualize the still life as a geometricized plane further strengthening their "perceptual and manual skills"; in other words, "the flat geometric design that students transferred to gridded slates were exercises en route to naturalistic representation. Kandinsky's method works inversely; the grid enables the student to coax geometry out of natural form" (Miller, 2019, p. 14).

Another pair of faculty members, Josef Albers (1888–1876) and Anni Albers (1899–1994), advocating for learning through systematic experiment, shared a pedagogical goal to instill in students a sophisticated and scientific understanding of the intrinsic characteristics of a material. Taking paper as an example, Josef Albers (1938/2017) asked students to test and observe "its properties of flexibility and rigidity, and its potentialities in tension and compression." This process emphasizing the hands-on experience with the basic law of a material strengthened students' "constructive thinking" (p. 116). Likewise, Anni Albers taught student to examine the physical properties of various fabric materials to grasp, for instance, their textural, light-reflecting, and sound-absorbing qualities. She impelled students to learn new machines and techniques in her weaving workshop and transformed ancient and women's craft into new forms of textile art like wall hangings and wall coverings suitable for industrial mass production and architectural purpose. Her weaving curriculum blending studies of materials, aesthetics, machines, and engineering yielded transdisciplinary textile art (Kim, 2022; Siebenbrodt & Schöbe, 2009).

3 An ADSTEAM Pedagogical Model

Inspired by the historical and Bauhaus philosophies of art, I devised an arts-driven STEAM pedagogical model, ADSTEAM, for short. This model illuminates the symbiotic relationships among STEAM subjects; the indispensable role STEM plays in the observation, appreciation, critique, and the making of art; and the goal of cultivating artistic habits uniting artistic and scientific thinking and doing. Clapp and Jimenez (2016) maintained that the "A" in an integrative STEAM experience stands for art learning, aesthetic education, and hands-on creativity. Following their advice, I developed the ADSTEAM model to provide students with opportunities to engage with selected arts disciplines, to nurture their aesthetic sensibilities, and to participate in creativity. The primary focus of ADSTEAM teaching is on art, STEM playing a vital role in broadening the appreciation of art and supporting studio instruction. The student's learning goal is to study and apply science, technology, engineering, or mathematics leading to a multifaceted comprehension of art and a well-rounded artmaking process. This model elevates the position and power of art in the STEAM curriculum and requires expertise in art, making it ideal for arts and humanities professors to adopt.

The ADSTEAM model features the following pedagogical strategies. In an art appreciation-oriented class, professors can update their curriculum by adding STEAM artworks and artists or highlighting the STEAM components in some master artworks to broaden students' inventory and knowledge of art. The assignments can compel students to explain the specific scientific and technological inventions, mathematical concepts, and engineering principles manifested in a theory, genre, or work of art. In a painting class, students interested in a still life of flowers can learn to observe and name the various types and parts of the flowers using scientific methods and terms. Whether the painting is abstract or realistic, the ADSTEAM approach enables students to create sensible and visually effective images of flower. Otherwise, professors can find inspiration in the Bauhaus curricular samples to enrich students' studio exercises (see Bayer et al., 1938/2017). In a studio critique assignment students can describe their STEAM content knowledge and STEAM infused creative process, precisely discussing individual work using professional, discipline-based language. To assess an ADSTEAM project, professors should provide evaluation criteria based on content knowledge, skills, and standards established in the disciplines incorporated in the project.

Exploiting these strategies, students can become cognizant of the intersections of STEAM and endeavor to identify, learn about, and then apply STEM knowledge pragmatically and immediately in an artistic context. In the long run, as more professors adopt the ADSTEAM pedagogy, akin to the transdisciplinary learning in the Bauhaus, students will be able to apply STEAM knowledge and skills to innovate new visual effects, forms of art, ways of manipulating materials, and visual communication schemes among others. Moreover, practicing the ADSTEAM strategies, professors likely help to demystify the stereotypes associated with artists and scientists as having disparate talents and training, bridge the academic gap between art and other academic subjects, and boost students' cognitive capacity by having them exercise transdisciplinary STEAM thinking.

4 ADSTEAM Learning Activities in an Undergraduate Theory of Art Class

Exemplified in this section is an ADSTEAM curriculum I have implemented in my undergraduate class on the theory of art since 2022. It is a senior-level art class, averaging 10 in each section, with the majority of students coming from the Department of Arts and Media and aspiring to become artists, designers, art teachers, or museum professionals. Employing the cyclical process of action research (Mertler, 2019), I reiterated the design, implementation, and evaluation of the curriculum. I learned from the 2022 pilot that a stand-alone lesson linking STEAM to the theory of art is not a gainful approach to cultivate in students a genuine and sustainable interest in ADSTEAM learning and in applying STEAM knowledge and skills in their creative process. Instead, references to the intersections and language of STEAM must sensibly permeate the entire curriculum in order for students to internalize the entrenched connections among STEAM subjects. Continuously modifying the curriculum, since 2023 I have instructed students to periodically read and discuss the historical, Bauhaus, and contemporary views of the intersections between art and STEM and taught the class following the ADSTEAM pedagogical model and strategies described in previous section.

From the beginning of the class, I nudged students to consider aspects of STEM in various theories, philosophies, and language of art as well as the intersections of STEAM knowledge and skills in a range of work by prominent artists throughout art history (e.g., Maria Sybilla Merian, Le Corbusier, Wassily Kandinsky, Maya Lin). As students grew familiar with thinking about the arts beyond the realms of arts and humanities and comfortable with the concept of STEAM in the theory of art class, they were ready to engage in a specialized ADSTEAM learning in the third module. Spanning about a quarter of the time in a semester, this module opens with reading and discussion assignments, contributing to further investigation of the codependence and mutuality among STEAM subjects since the early history of art, in the Bauhaus transdisciplinary philosophy, and for STEAM art and artists (e.g., Bullot, et al., 2017; Dietrich & Knieper, 2022; Freeland, 2001). Warmed up by this theoretical exploration, students then carried out an ADSTEAM creative project, applying STEAM knowledge to reexamine probing questions emerging from their favored theories of art and to explore STEAM aesthetics through a studio exercise. Students concluded the module with peer review and critique of their projects.

4.1 *Probing Question: What Makes a Work of Art Visually Appealing?*

To determine what STEAM knowledge to apply in the ADSTEAM creative project, I identified the recurring theories along with associated art forms, artworks, and artists students had chosen to explore in their discussion and writing assignments. I noticed that students were fascinated with the interplay of the theories of beauty and aesthetic emotion, visual perception, and neuro- and cognitive science. For example, students often traced back to the early class discussion of Freeland's (2001) inquiry into reasons that the red rose is perceived as beautiful. Referring to Kant's theory, Freeland stated: "Something about [a rose's] array of colours and textures prompts my mental faculties to feel that the object is 'right.'" She asserted that this "rightness" is crucial in Kant's perception of beauty and postulated that "we label an object beautiful because it promotes an internal harmony or 'free play' of our mental faculties; we call something 'beautiful' when it elicits this pleasure" (p. 12).

Next to Kant's theory, students were most influenced by Bell's (1913) legendary theory of significant form and avidly practiced formalist analysis of art. Bell proposed the following: "For a discussion of aesthetics it need be agreed only that forms arranged and combined according to certain unknown and mysterious laws do move us profoundly" and that "these moving combinations and arrangements [are called] significant form" (p. 227). Although the majority of students could clearly articulate Kant's and Bell's theories, they constantly questioned whether their theories perpetuate art as a subjective and enigmatic entity.

Inevitably, students continued to reckon with what makes certain arrangements of formal properties "right," emotionally "moving," aesthetically pleasing, or stimulating "free play" in their mental faculties. Recalling what they learned from respective studio classes, some students pointed out a prevalent concept of the rule of thirds. Those in the disciplines of photography, graphic design, and film were particularly indoctrinated into this aesthetic convention; however, they admitted a lack of knowledge of the history and STEM concepts behind the convention. Although some of them were aware that it is closely related to another compositional custom called the golden ratio, most were unaware that these aesthetic norms were associated with the mathematical concept of the Fibonacci sequence. They were familiar with the term because of Dan Brown's (2003) novel *The Da Vinci's Code,* and their curiosity intensified as they prepared to employ an ADSTEAM approach to unravel aesthetic norms shaped by the mathematics.

4.2 *The Fibonacci Sequence, the Golden Ratio, and the Rule of Thirds*

Detailed in *Master Fibonacci* (Allen, 2019), the now-called Fibonacci sequence was brought to the Western world by the Italian merchant Leonardo Pisano (ca. 1175–1250) also known as Fibonacci. In his 1228 revised book on mathematics called *Liber Abaci* (*The Book of Calculation*), Fibonacci posed a brainteaser remembered as "the rabbit problem," hypothesizing the growth pattern of rabbit pairs under idealized breeding conditions. The solution reveals a series of numbers—1, 1, 2, 3, 5, 8, 13, 21, 34, 55, 89, and so forth—corresponding to the number of rabbit pairs born in each month. The pattern in this series displays each number as the sum of the two previous numbers. This numerical phenomenon progressing indefinitely without repeats is the Fibonacci sequence.

As the numbers in the Fibonacci sequence increase after the first 10 ratios, the ratio of any two successive numbers remains close to 1.618. For example, the Fibonacci sequence grows to 55, 89, 144, 233, 377, and so forth. When a number is divided by its previous number, the ratio is close to 1.618 as shown below:

$$89/55 = 1.61818...$$
$$144/89 = 1.61797...$$
$$233/144 = 1.61880...$$
$$377/233 = 1.61802...$$

The approximate ratio of 1.618 is related to the golden ratio, derived by dividing a line consisting of two unequal sections of a and b (and a > b > 0) in such a way that the ratio of the whole line to the long section (a) is equal to the ratio of the long section to the short section (b). The result indicates that the ratio of a to b is an irrational number—1.6180339887 … —in which decimals proceed indefinitely and nonrecursively. It is symbolized by the Greek letter Phi (ϕ) as the golden ratio. This description can be expressed algebraically as follows (see Marples & Williams, 2022, for full mathematical work):

$$\frac{a+b}{a} = \frac{a}{b} = \phi = 1.61800339887...$$

The golden ratio, a famous formula in geometry, conceptualizes an aesthetically appealing proportion. As such, it is applied in myriad ways to create the so-called golden spiral and golden shapes of triangle, rectangle, and pentagram among others (Allen, 2019).

The rule of thirds is often compared to the golden ratio because they exhibit similar arithmetic and visual properties. The ratio of 2/3 in the rule of thirds is about 0.666, which is close to the inverse of the golden ratio (small and lowercase phi), which is about 0.618. In practice, artists adopt the golden ratio and the rule of thirds to divide a frame evenly into thirds horizontally and vertically and arrange objects accordingly to enhance visual flow. Caplin (2008) recounted that the rule of thirds was mentioned in a discourse on compositional principle in the 18th century, in which English artists recognized an eye-pleasing effect when a visual field is divided into three equal parts and the focal point or main subject is placed in the right or left third of the field, leaving the other 2/3 of field open. This preference persists in the disciplines of visual arts as affirmed by Adobe (n.d.), the company whose art and design software dominates the creative industry in the United States: “The rule of thirds is a composition guideline … While there are other forms of composition, the rule of thirds generally leads to compelling and well-composed shots” (para 4).

The rule of thirds, the golden ratio, and the Fibonacci sequence, having evolved respectively in various eras, have contributed to the development of STEAM knowledge, art, and aesthetics. The Fibonacci sequence, in particular, has numerous applications beyond calculating the rabbit population. It is employed in numerous fields: mathematics, influencing number theory, geometry, and more; natural science and computer science for modeling or describing various phenomena; the arts for observing and creating intriguing and pleasing visual, sound, or textual patterns; and finance for studying and analyzing stock trading (Allen, 2019; Marples & Williams, 2022).

4.3 *The ADSTEAM Assignment*

Sufficiently prepared and eager, students moved on to the studio ADSTEAM project. I provided the resources and instruction for them to complete three tasks:

1. Research and answer the following questions in mathematical terms, formulas, or graphics: What is the Fibonacci sequence? What is the golden ratio? How is the Fibonacci sequence related to the golden ratio?
2. Apply the Fibonacci sequence or the golden ratio to create an original visual work expressing a theme, feeling, idea, or purpose. Experiment with various ways these mathematical formulas can be used as visual compositional strategies to augment the principles of design—balance, movement and rhythm, variety, unity, harmony, pattern and repetition, and proportion—contributing to the desired intention of the visual work.
3. Write an artist statement explaining the Fibonacci sequence and the golden ratio, the intention of the visual work, and the creative process. In addition, the writing prompts included the following questions:
   - What visual effects and compositional dynamics (using vocabularies from the principles of design) did the Fibonacci sequence generate?
   - How did the use of the Fibonacci sequence or the golden ratio affect your ability to convey the intention of your visual work?
   - How has the studio ADSTEAM project affected your thinking about the aesthetic theories that you have explored in this course?
   - What insights do you have about the intersections of STEAM subjects and creating visual work supported by STEAM concepts and skills?

5 Students' ADSTEAM Projects

Sampled in this section are the students' ADSTEAM projects illuminated by their reflections on ADSTEAM knowledge, the creative process, and aesthetics. Notably, with abundant resources provided, the majority of the students correctly explained the Fibonacci sequence or the golden ratio in mathematical terms and formulas. They were able to illustrate the Fibonacci or golden shapes and spirals and used them as compositional templates to create visual work expressing a theme or emotion or serving a purpose. Their artist statements and peer critiques featured language and knowledge relevant to the discipline that constitute the STEAM field.

Take Student A's project as an example. Her intention was to create a poster promoting the ADSTEAM learning module by accentuating both the probing question—"What is art?"—and the Fibonacci sequence. An avid graphic design student, Student A claimed in her artist statement that she was familiar with the convention of using the golden ratio as a template to create harmony in design and "was previously introduced to the golden spiral but not the mathematical way of achieving the golden ratio" (personal communication, November 28, 2022). Thus, the project motivated her to probe this aesthetic convention. Student A explained in writing the Fibonacci sequence and the golden ratio. Using a digital means, she then created a series of golden rectangles based on the ratio of 1:1.618 (Figure 14.1). After such, she rendered arcs connecting each opposite corner of the square to form the golden spiral. Using the golden rectangles and spirals as a compositional template, she laid down the predominant geometric shapes.

Student A gradually and strategically filled in the shapes with prominent works of art, the probing question, and the Fibonacci numbers (Figure 14.2). She recounted:

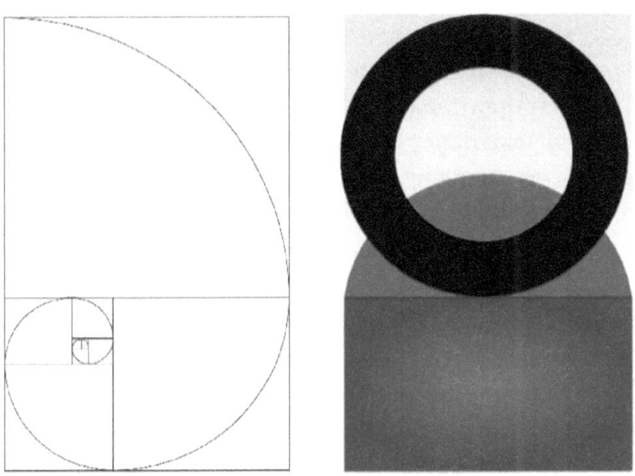

FIGURE 14.1 Student A's golden rectangles and spirals (left). The predominant geometric shapes placed over the golden rectangles (right), 2022

FIGURE 14.2 Student A's poster design in progress (left). Student A's final poster design (right), 2022

> The golden spiral/ratio helped me create balance and overall harmony in my layout. Despite the big circle at the top taking much of the space, the colors, and images I chose, [I was able to keep the] elements in line and in balance with the bottom half of the poster. (personal communication, November 28, 2022)

Reflecting on her ADSTEAM creative experience, Student A cited the benefits and limitations of applying STEM approach to artmaking:

> STEM knowledge and skills do help to create aesthetically pleasing a/symmetrical art that is proportional and balanced but can also possibly force an artist to create an experience that is limited and constrained with little individuality. However, knowing what the rules are and then breaking them is also an opportunity for a skillful artist to create art that is both unique and has form. (personal communication, November 28, 2022)

Student B had been working on a window painting depicting spring scenery for the school where she is a teacher assistant. Learning about the Fibonacci sequence, she decided to replan the composition to better express a pleasant, welcoming emotion associated with springtime. Initially, using mathematical and engineering tools, she drew a series of squares corresponding to the Fibonacci numbers—1^2, 1^2, 2^2, 3^2, 5^2, and 8^2—and sketched curves diagonally

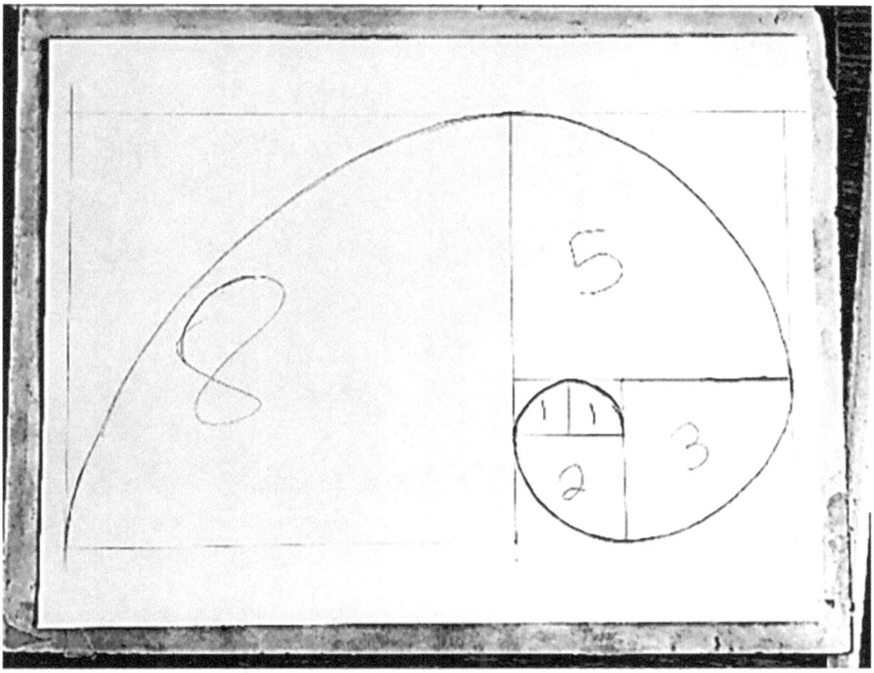

FIGURE 14.3 Student B's sketch of a Fibonacci spiral

connecting each corner of the square to form the Fibonacci spiral (Figure 14.3). This then became a compositional guide for her to rearrange and combine various objects and colors in an orderly, eye-catching, and gentle moving style. Her final work was a landscape gouache piece sized to match a Fibonacci rectangle of 8" × 13" (Figure 14.4). Critiquing her aesthetic sensibility, she asserted that the visual flow and color theory "capture the feeling or the warmth of spring":

> The diagonal–triangular shape of the road brings the viewer's eye toward the branches of the tree, down the trunk, and to the pond (the center point of the spiral). I put the cooler tones at the outside of the spiral and the warmer and brighter tones toward the center because our eyes are more attracted to the warmer tones. (personal communication, March 23, 2023)

Recalling ADSTEAM learning and the creative process, she concluded that mathematics and science were never separate from the art world by pointing out that "paints and solvents are chemistry. The way light works is physics. Our biology is wired for us to see and use objects for art. Science tells us *how* the world and the human body works, but art is our response and perceptions of the world" (personal communication, March 23, 2023).

FIGURE 14.4 Student B. *Fibonacci Koi Pond*. 8" × 13." Gouache on Stonehenge Print Paper. 2023

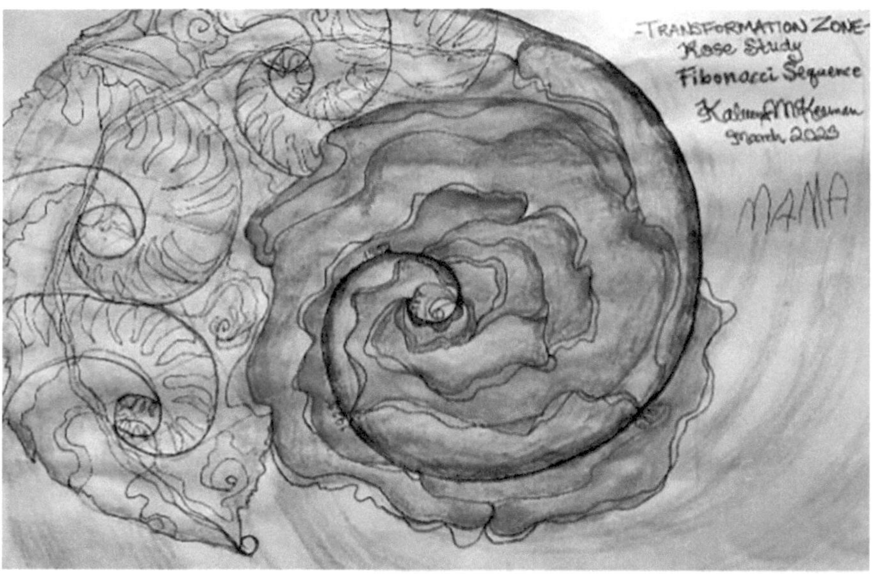

FIGURE 14.5 Student C. *Transformation Zone*. Watercolor, 2023

Student C found artistic inspiration and female, natural, and biological metaphors in the Fibonacci spiral. In her watercolor titled *Transformation Zone* (Figure 14.5), using multiple spirals as compositional guides and visual allegories, she expressed thoughts about the arts and humanities, the mathematical patterns, women's delicate biological conditions, the medical sign for "ferning," and the feminine symbol of the rose, representing beauty. Along the spiral, she penned down the birth years of the four female family members to represent cyclical time and childbirth. The excerpts from her artist statement, as collected below, show knowledge of the interconnections among STEAM subjects:

> The rose represents a rich array of religious elements and entities as well as purity, intuition, resurrection, rebirth, and many forms of love and beauty. The cyclical regeneration connects … many natural processes … like fractal patterns found in nature. The "Transformation Zone" [describes] the area in which irrelevant cell groups join and experience a state of constant alteration called "metaplas[tic] change" throughout the entirety of a woman's life within the cervix. This zone [having] a heightened level of fragility is well-represented by the blooming rose. "Ferning" [is] another medical term describing the crystals which form during the procedure to conduct a fertility test. (personal communication, April 3, 2023)

Explaining her position regarding ADSTEAM creativity, Student C asserted that scientific knowledge about the basic proportions of humans, animals, and nature; perspectives; and the visual perception of space are vital in art. If not following the knowledge, the depiction of the objects will be distorted and "childlike." More attuned to natural and life sciences, she thought that the Fibonacci spiral mirrored patterns in nature, such as the curled-up leaves, wrapped flower petals, and the cyclical characteristic of female life. Adopting the Fibonacci spiral as a compositional scheme, thus, was instinctually comforting as she affirmed that the spiral "naturally created a visual rhythm and balance within the piece. Additionally, the repetitive pattern through these exponential outward spirals draws the eye in and allows for a guided experience" for observing colors (personal communication, April 3, 2023). Student C noted that she was not highly aware of the connection between mathematics and art; however, her statement revealed an effort to grasp an interrelated natural and mathematical phenomenon of exponential growth, which is a mathematical language referring to population growth like the one analyzed in Fibonacci's rabbit problem.

Student C further emphasized the communicative and expressive elements of art for their ability to augment comprehension of STEM information. This resonates with one of the art theories examined in the class—the cognitive theory of art—in which art is perceived as something "enabling people to perceive, manipulate, or otherwise grapple with reality [and] something people use to enrich their world and their perceptions. [As] Dewey argued, … [it] can be a source of knowledge just as much as science" (Freeland, 2001, pp. 166–167). Student C speculated that the method of mixing art and mathematics may "evoke a subtle form of data mapping" and provoke the viewer to observe a work of art beyond its visible aesthetic elements to consider the scientific data and metaphors embedded in the work (personal communication, April 3, 2023).

Evidently, students A, B, and C strove to study the mathematical formula of the Fibonacci sequence and the golden ratio and visual diagrams derived from these formulas. Several students responded that comprehending the mathematics through textual explanation is difficult; however, when Students A and B illustrated and explained the Fibonacci sequence and the golden ratio in a graphic and colorful way, their peers found themselves able to engage in the discussion of the mathematics using proper STEAM language and of the way the Fibonacci sequence can be adopted and varied to attain certain visual effects provoking aesthetic emotion. Equipped with the ADSTEAM creative process, several students admitted that they had become more aware of the principles of design and capable of evaluating and critiquing one another's visual work in an analytical and cogent—as opposed to intuitive and impressionist—manner. Another mutually agreed sentiment emerging in every class was that ADSTEAM learning enables students to concretely appreciate aesthetic theories like significant form, formalism, expressionist theory, and cognitive theory. Student A stated:

> [STEAM] can help inform a critique of a piece of art where the formalism theory can be applied. This [ADSTEAM] exercise makes the evaluation of other artworks easier to do. It brings math and science into the art world and helps us reevaluate art and the theory of beauty. (personal communication, November 28, 2022)

And Student D followed:

> I feel I am searching for that spiraled composition within different works of art… This makes viewing art much more interesting. It is no longer about viewing it in regard to "l'art pour l'art" but now in analyzing the [STEAM-infused] creative process. (personal communication, November 30, 2022)

6 Final Reflection

A literature review of the intersections of STEAM subjects and the Bauhaus aesthetic philosophy has encouraged me to devise and implement the ADSTEAM pedagogical model. In this chapter I detailed three characteristics of the model applied to my undergraduate theory of art class. First, I emphasized the symbiotic relationships among STEAM subjects by introducing the STEAM paradigm and art developed in the Bauhaus school and by asking students to apply the Fibonacci mathematical sequence along with artistic and engineering concepts and skills to create original visual works. Second, I reinvigorated students' belief in the indispensable role STEM plays in the observation, appreciation, critique, and making of art when they realized the mathematical work behind the visual compositional convention they had been practicing and understood that they were using STEM knowledge and language to critique one another's visual work. Third, students have demonstrated artistic habits uniting artistic and scientific thinking and doing through the ADSTEAM creative process.

My experimentation with the ADSTEAM pedagogy reveals that STEAM knowledge can be featured in arts and humanities classes to broaden students' appreciation of art beyond the discipline-based content and mode. STEAM principles can be infused into studio art activities in a challenging way while augmenting students' creative thinking, quality of artworks, and potential in generating innovative STEAM aesthetics. This clearly resonates with Clapp and Jimenez's (2016) concept of the "A" in STEAM as arts learning, aesthetic education, and creativity. The students attested that the ADSTEAM projects fostered a transdisciplinary way of knowing and artmaking when they moved beyond single discipline-based approach to generate knowledge of art and original visual works. The Fibonacci mathematical concept empowered them to comprehend and visually respond to Bell's (1913) iconic yet mysterious aesthetic of significant form.

The ADSTEAM pedagogical model can pave the way for professors to transform canonical content knowledge and instructional conventions in the arts and the humanities. For example, my students pointed out that they had learned the compositional standards in graphic design based on the golden ratio and the rule of thirds but not the mathematical work behind them. Advocating for ADSTEAM learning, professors need not be shy away from discussing with students or asking students to explain and apply mathematical and scientific principles and processes in arts and humanities classes. They can impel students to describe STEAM elements in the renowned works and theories of art. Studio instruction and critique are no longer restricted to artistic matter but can include questions about ways STEM is integrated into the creative process. Ultimately, the ADSTEAM pedagogical model can contribute to a holistic arts and humanities education, expanding the epistemology and ontology of art.

References

Adobe. (n.d.). *How to use, and break, the rule of thirds.* https://www.adobe.com/creativecloud/photography/discover/rule-of-thirds.html

Albers, J. (2017). Preliminary course: Albers; concerning fundamental design. In H. Bayer, W. Gropius, & I. Gropius (Eds.), *Bauhaus, 1919–1928* (pp. 116–123). Museum of Modern Art. (Original work published 1938)

Allen, S. (2019). *Master Fibonacci: The man who changed math.* Fibonacci Inc.

Areljung, S. (2023). Five ways of integrating arts and science: A framework for planning and analyzing arts–science education in early childhood. *Studies in Art Education, 64*(1), 9–22. https://doi.org/10.1080/00393541.2022.2154522

Barr, A. H., Jr. (2017). Preface. In H. Bayer, W. Gropius, & I. Gropius (Eds.), *Bauhaus, 1919–1928* (pp. 1–10). Museum of Modern Art. (Original work published 1938)

Bayer, H., Gropius, W., & Gropius, I. (2017). *Bauhaus, 1919–1928.* Museum of Modern Art. (Original work published 1938) https://doi.org/10.2307/1574173

Bell, C. (1913). Post-impressionism and æsthetics. *The Burlington Magazine for Connoisseurs, 22*(118), 226–230.

Brown, D. (2006). *The Da Vinci code.* Doubleday.

Bullot, N. J., Seeley, W. P., & Davies, S. (2017). Art and science: A philosophical sketch of their historical complexity and codependence. *The Journal of Aesthetics and Art Criticism, 75*(4), 453–463. https://doi.org/10.1111/jaac.12398. Top of FormBottom of Form

Burnard, P., & Colucci-Gray, L. (Eds.). (2020). *Why science and art creativities matter: (Re)configuring STEAM for future-making education.* Brill. https://doi.org/10.1163/9789004421585

Caplin, S. (2008). *Art and design in Photoshop.* Routledge. https://doi.org/10.4324/9780080928289

Clapp, E. P., & Jimenez, R. L. (2016). Implementing STEAM in maker-centered learning. *Psychology of Aesthetics, Creativity, and the Arts, 10*(4), 481–491. https://doi.org/10.1037/aca0000066

Cohen, E., & Lloyd, S. (2014). Disciplinary evolution and the rise of the transdiscipline. *Informing Science: The International Journal of an Emerging Transdiscipline, 17*, 189–215. https://doi.org/10.28945/2045

Dietrich, P., & Knieper, T. (2022). (Neuro)aesthetics: Beauty, ugliness, and ethics. *PsyCh Journal, 11*(5), 619–627. https://doi.org/10.1002/pchj.478

Freeland, C. (2001). *But is it art?* Oxford University Press.

Herro, D., Quigley, C., & Cian, H. (2019). The challenges of STEAM instruction: Lessons from the field. *Action in Teacher Education, 41*(2), 172–190. https://doi.org/10.1080/01626620.2018.1551159

Hunter-Doniger, T., & Walkup, N. (Eds.). (2024). *STEAM Education: An interdisciplinary look at art in the curriculum.* National Art Education Association.

Kim, H. (2022). Re-framing Anni Albers and Bauhaus. *International Journal of Art & Design Education, 41*(3), 414–426. https://doi.org/10.1111/jade.12423

Lai, A. (2024, April 05). *Reviewing and reimagining research on STEAM education.* [Slide presentation]. The 2024 National Art Education Association Convention, Minneapolis, Minnesota, United States.

Liao, C. (2016). From interdisciplinary to transdisciplinary: An arts-integrated approach to STEAM education. *Art Education, 69*(6), 44–49. https://doi.org/10.1080/00043125.2016.1224873

Marples, C. R., & Williams, P. M. (2022). The golden ratio in nature: A tour across length scales. *Symmetry, 14*(10), 2059. https://doi.org/10.3390/sym14102059

Marshall, J. (2014). Transdisciplinarity and art integration: Toward a new understanding of art-based learning across the curriculum. *Studies in Art Education, 55*(2), 104–127. https://doi.org/10.1080/00393541.2014.11518922

Mertler, C. A. (2019). *Action research: Improving schools and empowering educators* (6th ed.). Sage.

Miller, J. A. (2019). Elementary school. In E. Lupton & J. A. Miller (Eds.), *The ABC's of [triangle, square, circle]: The Bauhaus and design theory.* (pp. 10–27). Princeton Architectural Press.

Pirrie, A. (2020). Where science ends, art begins? Critical perspectives on the development of STEAM in the new climatic regime. In P. Burnard & L. Colucci-Gray (Eds.), *Why science and art creativities matter: (Re)configuring STEAM for future-making education* (pp. 19–34). Brill. https://doi.org/10.1163/9789004421585_003

Siebenbrodt, M., & Schöbe, L. (2009). *Bauhaus.* Parkstone International.

PART 4

Remapping Teacher Education through STEAM *Practices*

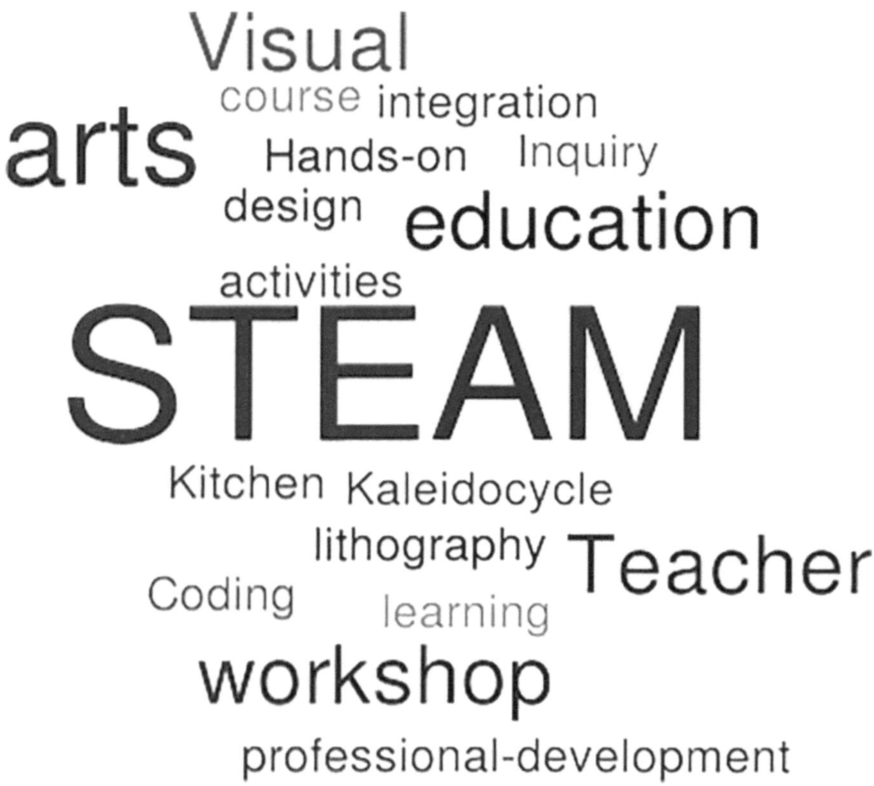
Visual
course integration
arts
Hands-on Inquiry
design education
activities
STEAM
Kitchen Kaleidocycle
lithography Teacher
Coding learning
workshop
professional-development

CHAPTER 15

STEAM-Oriented Instructional Design in Visual Arts Teacher Education in Türkiye

Suzan Duygu Bedir Erişti and Yeliz Erdoğan

Abstract

In this chapter we explain the process of planning STEAM-oriented instructional activities in visual arts education, sharing our experiences with Art Teaching Practices, a course we initially offered in fall 2020. We highlight several of our instructional strategies, including 5E4C, daily-life and context-based STEAM learning, the maker movement, and technology-enhanced STEAM projects. We also recount a range of activities carried out by our students, who were teacher candidates preparing themselves for STEAM education; these included analyzing STEAM examples in film and social media, connecting Türkiye's visual arts competencies to STEAM projects, participating in STEAM workshops to learn with and from international peers and experts in STEAM education, and making STEAM-oriented art projects, such as kaleidocycles, smart space design using a coding program, and kitchen lithography.

Keywords

STEAM course design – visual arts teacher education in Türkiye – kaleidocycle – coding – kitchen lithography

1 Introduction

In our literature review of STEAM education in Türkiye, we found almost no studies of STEAM-oriented art education courses or their implementation in the teacher education process; in fact, numerous scholars have concurred with our finding (Atalay, 2019; Azkın, 2019; Bozkurt, 2019; Erdönmez, 2019; Hallaç, 2019; Helvacı, 2019; Kolsuz, 2018; Mercan, 2019; Uştu, 2019). Ministry of National Education [MoNE] (2018) standards for STEAM education in Türkiye have focused on interacting, exchanging, investigating, creating, presenting, and developing. Ministry projects, including STEAM, are supported with

 | DOI:10.1163/9789004714748_016

consultancy, training, workshops, and seminars to promote a multidisciplinary approach. Based on our observation, arts education content and projects are often omitted from these trainings. Consequently, despite the Ministry's support of STEAM education, the majority of the project-based curriculum influenced by STEAM education models at private and public schools, in our opinion, lack "A," that is, an understanding of how art education can foster creativity. Thus, these so-called multidisciplinary approaches in STEAM education focus only on teaching visual qualities and the use of elements of art in the presentation of STEAM topics. As the MoNE moves to include in its educational standards innovative competencies to solve the problems of daily life with interdisciplinary projects, we strongly recommend that the STEAM curriculum be structured to include all disciplines to foster students' innovative competencies.

In this chapter we recount our experience in planning STEAM-oriented instructional activities in visual arts education. Two back-to-back special issues on STEAM education published by *Art Education* emerged as an important resource, guiding us in planning STEAM activities (see Rolling, 2016a, 2016b). We learned that STEAM is an educational model that expands the ways that academic subjects like science, technology, engineering, the arts, and mathematics are structured in the design of the integrated curriculum. Another source guiding us in planning STEAM instruction was a conference called STEAM Education and Visual Arts and a particular workshop led by Dr. Yichien Cooper in Ankara, Türkiye, in 2018. As researchers we found these resources vital for our making sense of the position of art in STEAM.

We recognize several key takeaways influencing our thinking about STEAM education: Art can be a significant stakeholder or component on its own; that is, art can be the central inquiry in STEAM education. Art can exist in STEAM education as an subject auxiliary to other disciplines. The dimension of STEAM that encourages learners to develop their innovative and creative perspectives is exactly what visual arts education needs. STEAM education provides opportunities to foster students' 21st-century skills: learning by doing, developing a critical perspective, and establishing relational contexts (Institute for Arts Integration and STEAM, 2019; Maeda, 2013; Yakman, 2010). Dr. Cooper (2018) explained that engage, explain, explore, elaborate, and evaluate (5E) are the cognitive practices embedded in the art education experience. Within these practices, communication, creativity, cooperation, and critical thinking (4C) can be strengthened. This 5E4C, often accentuated in STEAM education, can foster learning that can be applied to real-world situations. In essence, the interaction of science and art in STEAM education allows the creation of art to be meaningfully integrated into the daily STEM context.

2 STEAM Education Preparation Process and Instructional Activities

As we aimed to improve the integration of STEAM into visual arts teacher education, we identified our course of action by analyzing the samples, mistakes, ideas, diverse perspectives, application methods, and strategies through a careful literature review as noted above. Consultation with art and education experts produced ideas to help identify the teaching content and determine whether we would implement the teacher training program by incorporating an existing or new lesson. After considering the number of students (teacher candidates) and the grade level in which they teach, we opened a course titled Art Teaching Practices capped at 12 students and structured to enrich understanding of and experience with the STEAM-oriented teaching process.

In our course we designed STEAM-oriented instructional activities to include content reflecting real-life experiences and a learning process focusing on the 5E4C competencies. Our goals were to instill in teacher candidates' aptitudes to create STEAM-oriented teaching plans and evaluations. Our STEAM approaches emphasized art integration, design thinking, learning with fun, and project-based learning. We implemented learning activities incorporating technology, mobile applications, art practices, and daily-life events, and provided teacher candidates with a student-centered, open learning environment, enabling them to design and experiment with their own STEAM-oriented art projects and lessons. Some students included mobile apps, augmented reality-based apps, game-based activities, and kitchen lithography. To expand their appreciation of a multidisciplinary approach to designing STEAM lessons and the maker movement, our course included interaction and discussions with international peers and experts on STEAM education, robotic coding and smart space design based on it, and stop motion animation production. Thus, we were able to equip teacher candidates with various teaching methods, techniques, and materials that they could use in STEAM education.

2.1 *Films and Videos*

To show scientists applying STEAM disciplines effectively amid social problems in the larger world, we showed the film *Hidden Figures* (Melfi, 2016), based on a true story of three Black women who contributed to the international space race. Watching this movie, teacher candidates grasped the holistic structure and relationality of philosophically different disciplines in STEAM. They critiqued the processes used to integrate multiple disciplines needed to develop space technology, concluding that many disciplines must be integrated to produce solutions to the problems of modern life. When analyzing the process portrayed in the film, students focused on policies, technology, sociocultural environment, science, racial issues, the women's rights movement, and recent breakthroughs in science.

Selecting social media videos on current developments in science and technology to connect with students' prior knowledge in science and technology, we introduced the work of Baris Özcan, a storyteller on social media, who creates educational videos to address science and technology inquiries as foundations for both teachers and students. For example, he explains what technological achievements have been accomplished today and what predicted breakthroughs may await us in the future in his videos titled *What Will the World Be Like in 2050? There Is No Such Person!* and *A Digital City* (Özcan, 2019a, 2019b, 2019c). As a class we discussed the rapid progress of technological developments; however, as we watched these rapid developments with excitement, we also admitted that we could not catch up with many of the groundbreaking scientific phenomena. After consideration, we realized that our education systems should make radical improvements to provide teachers and students with information related to progressive developments in science and technology.

We then showed *The Wind Rises*, an animated film by Hayao Miyazaki (2013), which we believe reflects the concept of STEAM in daily context. The film depicts the realization of an aircraft through the integration of STEAM disciplines; that is, imagination, creativity, the engineering design process, and the prototype production–testing–reproduction cycle are integral to the conceptualization and construction of the aircraft. In our view this resembles the real-life application of the STEAM education process. The relevance of STEAM in daily life was another important matter we wanted the students to grasp. They were most impressed that a character conceived a design idea inspired by a meal she ate.

2.2 *Maker Movement*

We adopted the characteristics of the maker movement into our course design and STEAM-oriented education process. Dale Dougherty (2012), an influential figure in the emergence and naming of the idea, a computer scientist, founder of *Make* Magazine, president of the Make Community, and creator of Maker Faire, asserted that everyone is a maker; people who cook for their families, knit, and garden are all "makers." Dougherty identified the maker movement with technology. This growing movement has been embraced by hobbyists, mechanics, engineers, computer experts, and artists dedicated to designing and building by combining materials creatively with fun and a philosophy of utilitarianism. We understand that the maker movement is generally an out-of-school phenomenon involving adult participants or makers, whereas STEM or STEAM education involves teaching students in a formal education setting to increase their participation in the engineering and design process and to improve their practices (Martin, 2015). Still, we see the potential of applying the maker movement to encourage students to see STEAM learning as fun and STEAM projects as practical in daily life.

Following Dougherty, we concluded that the maker movement can support arts education. We associated the basic principles of STEAM and the maker event effectively in the process. While designing the activities needed to find a solution for a STEAM-based problem, we encouraged teacher candidates to explore their opportunities. We observed that students improved their problem-solving skills and creativity during these activities. Our observation of activities included the interaction of STEAM and the maker movement during the process.

We see the traits of the maker and STEAM-infused creativity often in our daily lives, even applying them unawares. For example, we can see STEAM in the solutions we find to small problems we encounter daily. In our youth all the neighborhood children built their daily-life needs from materials found in nature—mud and stone. Children made everything while playing, pretending, using their imaginations to construct and play with creations with their friends. If a wheel broke, we fixed it with mud; and we used our imaginations to develop designs for objects not found in our homes. We built our own worlds with what we now consider to be STEAM play. Thus, in our opinion, STEAM can be found in everything: the games we play, the movies we watch, and the books we read.

2.3 *Coding, Games, and Mobile Technology-Based STEAM Learning*

We decided to employ a coding-based lesson to integrate the STEAM-oriented learning process into visual arts education. To increase our qualifications as instructors, we participated in a series of coding trainings and seminars. We chose Arduino, an easy-to-use, open-source electronic platform, and Mblock, a tool that integrates the Arduino coding syntax. With the help of the Arduino coding language, we can activate a motor, switch a light on and off, work on a sensor, or post something online. Arduino has been the brain of thousands of projects over time—from everyday objects to complex scientific tools to artistic works (*What is Arduino?* 2018). We chose Arduino and similar robotic coding tools, popular in the maker movement, because they can be applied to early age groups, affect 21st-century skills competences, and are easily programmable. Through robotic coding education individuals can gain interdisciplinary experience and learn algorithmic thinking skills as well as creative thinking and problem-solving skills (Sönmez & Şahinkayası, 2021).

At the beginning of the course preparation process, we independently and collectively participated in coding-related training, workshops, and activities. As observers we investigated the robotics and coding curricula of Türkiye's Science and Art Centers, which provide gifted youngsters with an education. Communicating with Science and Art Center teachers and asking them about their course and their experiences with project procedures, we gained technical skills and refined our viewpoint on developing a STEAM-oriented curriculum.

Meanwhile, the COVID-19 pandemic erupted worldwide, leading to the increased use of mobile educational technology. Thus, we shifted our focus to examine mobile apps useful in STEAM training for teacher candidates. For the participants and next-generation teachers, using mobile apps in STEAM education was very meaningful, modern, motivating, and related to daily life. We discovered mobile apps for coding and the production of artistic content as well as interesting apps and games related to types of augmented reality, becoming more selective about mobile apps and how we would use them in our lessons for educational purposes. Continuing to search for mobile apps before and during our preparation for the course, we have researched and examined nearly 50 apps, experiencing at least 20 of them via mobile devices. Based on our own experiences, we created the main list of the apps we thought would be useful for teacher candidates.

2.4 *Kaleidocycle Construction*

Apart from mobile apps, game-based activities also captured our attention. After all, sometimes we can teach more easily through simple games. We found kaleidocycle-making an interesting game-based activity for teaching STEAM content. During the preparation process for the course, we examined examples of and templates for kaleidocycle, experimented with different types of paper, and used ready-made patterns. Because of our logic of learning before teaching, we made the first attempts to draw our own kaleidocycles. In our experiments we saw that the thickness of the paper affected the transformation of the form, that the drawing of the template requires mathematical skills, and that the direction of the design and the principle of complementarity should be considered. Impressive kaleidocycles can be made with even simple drawings and designs. We decided that making a kaleidocycle would be fun for teacher candidates because it has a visual structure that gives the feeling of popular games like Pokemon and Beyblade and offers new designs in every spin.

The kaleidocycle comprises six equilateral triangles that move in a single direction and are made of folded and glued paper. The types of kaleidocycle depend on their mobility capabilities. Schonke and Fried (2019), researchers at Okinawa Graduate University of Science and Technology, suggested that

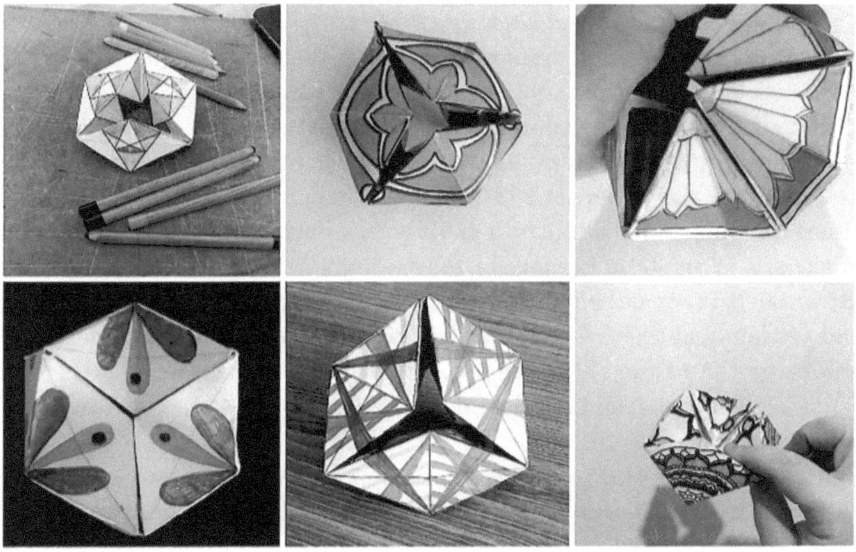

FIGURE 15.1 Kaleidocycle designs of the lecturers and teacher candidates (left to right): Top row: Yeliz Erdoğan, Yeliz Erdoğan, Zehra Dede. Middle row: Esra Kuş, Gamze Gezer, Sena Sevinç. Bottom row: Sümeyye Güneş, Ela Kantarcı, Birsen Çelik, 2020, Eskişehir, Türkiye.

the Mobius kaleidocycle could advance synthetic chemistry and robotics. When creating a kaleidocycle, the two ends of the triangle chain are joined, ensuring that the angle between consecutive joints is exactly 90 degrees. This relationship ensures that the kaleidocycle rotates with perfect triple symmetry inside out. The kaleidocycle permits movement in a variety of directions. At the intersection of science, art, and mathematics, its movement patterns capture creativity. As shown in Figure 15.1, the ring connections, which consist of joints and solid geometric shapes, are reminiscent of a flower bud that repeatedly blooms (*Mobius Kaleidocycles*, 2018; Schönke & Fried, 2019). Most of the teacher candidates struggled to build the kaleidocycle, trying multiple times before they could successfully construct one. The kaleidocycle, which we incorporated into our course as a recreational activity integrating STEAM disciplines, has provided new insights that have led teacher candidates to mathematical analysis and creative design concepts.

We continued to incorporate film and video, mobile applications, robotic coding, game-based activities, animation, and workshops into our STEAM-focused teaching activities. Our aim in the exercises was to demonstrate to the teacher candidates what they are capable of by bringing together many aspects of daily life.

2.5 *Students' STEAM Activities and Projects*

In the Turkish education system the visual arts curriculum includes three learning areas: visual communication and form, cultural heritage, and art criticism and aesthetics. The competencies in each area vary according to grade level. The term "competency," also viewed as "objective," denotes what students can do, know, and understand (Donnelly & Fitzmaurice, 2005). For example, visual communication at the fifth-grade level entails "using digital technology in visual arts"; cultural heritage at the sixth-grade level involves "examining the artifacts of Anatolian civilizations through museums"; and art criticism and aesthetics at the seventh-grade level focus on "explaining experiences of art practice" (Ministry of Education, 2018). As a requirement we asked teacher candidates to design and evaluate their STEAM activities to reflect the visual arts competencies.

Because movies and social media are popular and important in learners' daily lives in the 21st century, we asked teacher candidates to find movies that gave them ideas about STEAM in real-life contexts and STEAM activity videos from social media. While watching and considering the theoretical narratives in movies and social media videos, teacher candidates were to point out the visual arts competencies presented there and discuss how they planned to design STEAM activities in visual arts classes.

The teacher candidates then worked on a video to demonstrate and explain a STEAM activity. We used a systematic approach, extending this STEAM instructional video activity across three weeks. Afterward, we facilitated teacher candidates' discussion of competencies compatible with the activities and what ideas for alternative activities could be produced in line with those competencies. While engaging in their STEAM activities, they learned by doing, critically evaluated the activities, and opened the possibility of including STEAM contexts in the teaching process. Overall in this lesson, we emphasized creativity, originality, and association with daily life, establishing the relationship between the visual arts lesson and STEAM.

We thought prompting teacher candidates to apply the concept of STEAM education to an activity connected with their real-life experience was important. To this end, they participated in an event—smart space design—in which they applied what they learned in the coding session to the space model they imagined. We wanted them to complete their designs by associating them with daily life and encouraged them to engage in brainstorming, focused group discussions, and idea cloud activities. As a result, they combined learning robotic coding and the visual arts to visualize objects and living spaces usable in daily life (see Figure 15.2). For instance, one teacher candidate who constructed her own house included a sensor and LED inside the refrigerator model so that the

FIGURE 15.2 Model designs by Buse Buket Turan, Ela Kantarcı, Sena Sevinç, Esra Kuş (left to right), 2020, Eskişehir, Türkiye

light goes on and off when the door is opened and closed. Several students similarly constructed a door light for their homes. They could apply these inventions every day in smart-space designs. In a second example a student who enjoyed reading in a rocking chair created and incorporated one into her smart space model. Studying robotic code for the first time in their lives and pairing it with model designs, these teacher candidates were able to attain basic manufacturing success.

The maker movement is identical to and supportive of STEAM education and an important trend in STEAM and visual arts education. We exemplified the STEAM point of view and showed the maker movement in the context of the arts through Dr. Tezcan Bahar's (2019) Kitchen Lithography workshop. The interaction of art and other disciplines in this workshop opened a new path for teacher candidates by showing the common points of the maker movement and allowing them to reevaluate their knowledge from a STEAM perspective.

Kitchen Lithography is an application based on creating products with the effect of lithography. It is an alternate, nonhazardous method of printmaking that does not involve toxic chemicals like acid. Tin foil, acetate, a lithography pen or oil pastels, printmaking ink or oil paint, powder, paper, Coca-Cola, a deep bowl, brush, natural sponge, and roller are used for kitchen lithography. These materials are conveniently accessible at home and can be found in the kitchen. After drawing a design on the foil with an oil pen, it should be placed in a Coke-filled container. The acid in the Coke has a corrosive effect on aluminum foil. In the subsequent stage the foil remains in the container until it is ready for printing. At this stage a natural sponge can be used to apply paint to the foil. A spoon can be used to apply pressure on the foil to transfer paint to a sheet of paper. The procedure involves a chemical reaction between oil and water, allowing students to transfer the design to paper. The chemical processes and materials of the technique incorporate elements of science, art, design, and engineering (Bahar, 2019).

Teacher candidates witnessed the interaction of multidisciplinary domains in structuring an artistic practice through the lesson in kitchen lithography, which directly aligns with STEAM education. The materials used for kitchen lithography reflected one of our reasons for preferring sustainable alternative printmaking practices over toxic materials that are harmful to the environment and human health. The creative process enabled students to gain a new perspective on their kitchens, where they spend a significant part of their time in daily life. As these teacher candidates created their prints (see Figure 15.3) in their kitchens, explaining how STEAM learning can be realized in a daily context became easier. In addition, the entire implementation process provided a new viewpoint on the technical and chemical properties of kitchen products. Teacher candidates structured their design processes by observing one another's production phases. Everyone proposed a solution to the challenges caused by material or chemical reactions in the process. They benefited considerably from the hands-on, creativity-based critical inquiry embedded in kitchen lithography.

FIGURE 15.3 Kitchen lithography: printmaking with drawing on aluminum foil, emulsion by Buse Buket Turan and Sena Sevinç, 2020, Eskişehir, Türkiye

Aside from hands-on STEAM activities, we offered several events for teacher candidates to brainstorm the approaches to STEAM education with international peers and experts; for example, we implemented a workshop hosted by Dr. Yichien Cooper at Washington State University Tri-Cities. Dr. Cooper and our team designed interactive and integrated art lessons then brought her students and our students together via Zoom. Students had the opportunity to see, evaluate, and share ideas about STEAM education from the perspectives and approaches used in the United States and Türkiye. Students from both sides of the globe talked about STEAM-focused activities and lesson plans, raised questions to one another, and commented on one another's lesson ideas or plans. Our students' comments on STEAM education focused on studio activity because they did not have enough time to include other components into their lesson ideas or plans. By comparison, the stages in STEAM curriculum design were included by Dr. Cooper and her students, so our students could observe necessary planning stages and evaluation processes to develop well-rounded STEAM-oriented lessons. We held another workshop for teacher candidates in Türkiye at the end of the semester to meet with Dr. Cooper, who explained her holistic perspectives on STEAM education, how to approach the course design process, and what to consider when preparing a lesson plan. She then asked our teacher candidates to develop an activity plan following similar logic. Teacher candidates, who shared their ideas and practices with Dr. Cooper through our Facebook group, were influenced by this approach and used it in preparing their lesson plans.

Another event was organized with specialist Elif Karademir, a robotics-coding teacher at the Science and Art Center in Türkiye, which educates gifted elementary school students. She shared her experiences with our students and demonstrated what teacher candidates can do. Using her personal experiences to address teacher candidates' concerns about technical deficiencies and inequality of opportunity, she noted the factors she considered when developing her lesson plans: daily life, creativity, and teamwork.

In our final workshop we talked with Dr. Ayvaz Tunç about the relationship between STEAM education and animation. Because animation is an interdisciplinary branch of art, it is distinguished as a versatile field associated with creativity in the context of STEAM. Dr. Tunç emphasized the need for thought and a knowledge base that are interdisciplinary in nature when describing the logic of animation and how it may be done during the workshop. He discussed the use of the fundamentals of physics, chemistry, and mathematics in order to move objects, figures, and drawings as well as how to create cell animation with mobile apps. After a one- or two-hour session, the teacher candidates, who had been creating animation all week, came up with amusing drawings. As part of the workshop activities, they analyzed their animation-making experiences and products, posing questions on how the movement of an object influenced the scene, how the animation related to other fields, how it was put together, and how technology was employed in general.

3 Conclusion

Reflecting on our experience in planning STEAM-oriented instructional activities in visual arts education, we observed that the learning environment, the STEAM-based cooperative dimension, and friendly communication between the course lecturers and students all positively strengthened the educational process. The learning environment enriched by various STEAM activities and interactions with international peers and STEAM education experts made learning interesting, lasting, and flexible for teacher candidates. Interviews with them revealed that communication with faculty members enhanced their interest in the lesson, level of activity, and belief that they would succeed. In our course "learning with fun" was a concept integrated with STEAM, positively affecting teacher candidates' attitudes toward STEAM learning. STEAM education has been seen as an approach that makes education and training fun for teacher candidates. Influenced by an enjoyable educational atmosphere, the course lecturers and the teacher candidates developed amicable relationships when communicating diverse views about STEAM education and plans, ultimately contributing to effective student learning.

In sum, we highlighted a range of approaches deemed vital in carrying out our STEAM-oriented instructional design in visual arts teacher education. Our STEAM education approaches included a holistic strategy to integrate multiple disciplines; prepared students to connect their lesson plans to the visual arts competencies; prompted students to situate STEAM learning in daily life and familiar sites; focused on design-based inquiry, problem-solving, and hands-on learning models; applied the 5E (engage, explore, explain, elaborate, evaluate) or 7E (elicit, engage, explore, explain, elaborate, evaluate, extend) learning steps; integrated the engineering design process into artistic practices; and fostered 21st-century skills. Combining these approaches to structure a course was a challenging task, requiring experience and time. Thus, we believe that although one semester has provided teacher candidates with a certain level of competence to prepare for a foundation for STEAM education, understand its philosophy, select an idea to create a lesson plan, and obtain competence in STEAM education, they require more time to learn and experience STEAM education to effectively teach their own STEAM lessons. In other words, individuals must develop and evolve over time by continuously participating in STEAM courses and training, through which they can sharpen their STEAM teaching methods and techniques, their application of new technologies and technological methods, and their course designs.

References

Atalay, M. (2019). *Meslek liselerinde steam etkinliklerinin matematik dersine yönelik tutum ve başarıya etkisi [The effect of STEAM activities in vocational high schools on attitudes toward mathematics lessons and success]* [Master's thesis]. Erciyes University.

Azkın, Z. (2019). *Steam (fen-teknoloji-mühendislik-sanat-matematik) uygulamalarının* öğrencilerin *sanata yönelik tutumlarına, steam anlayışlarına ve mesleki ilgilerine etkisinin incelenmesi [Examining the effects of STEAM (science-technology-engineering-art-mathematics) applications on students' attitudes toward art, their understanding of STEAM, and their professional interests]* [Master's thesis]. Karamanoğlu Mehmetbey University.

Bahar, T. (2019). As an alternative printmaking technique waterless lithography. *The Journal of Social Science, 6*(33), 115–130. http://dx.doi.org/10.16990/SOBIDER.4756

Bozkurt, Y. (2019). *Steam etkinlikleri ile 7. sınıf* öğrenc*ilerinin başarı ve tutumlarındaki değişimin cinsiyete göre analizi* [Change in achievement and attitudes of *seven*th-grade students through STEAM activities: Analysis by gender] [Master's thesis]. Erciyes University.

Cooper, Y. (2018, May). *STEAM education and visual arts.* [Paper presentation]. STEAM Education and Visual Arts Conference, Nesibe Aydın Schools, Ankara, Türkiye.

Donnelly, R., & Fitzmaurice, M. (2005). Designing modules for learning. In G. O'Neill, S. Moore, and B. McMullin (Eds.). *Emerging issues in the practice of university learning and teaching* (pp. 99–110). All Ireland Society for Higher Education.

Dougherty, D. (2012). The maker movement. *Innovations: Technology, governance, globalization, 7*(3), 11–14. https://doi.org/10.1162/INOV_a_00135

Erdönmez, İ. (2019). Özel *yetenekli* öğrencilerin *coğrafya eğitiminde scamper tekniği ile steam uygulamaları [STEAM applications of talented students in geography education with Scamper technique]* [Master's thesis]. Gazi University.

Hallaç, S. (2019). *Disiplinlerüstü bir steam yaklaşımı ile hazırlanmış* öğretim *programının* öğrencilerin *fizik kavramlarını* öğrenmelerine, *bilime karşı tutumlarına, steam tutumlarına ve kariyer seçimlerine etkisinin incelenmesi [Investigating the effect of a superior STEAM approach to learning the physics on attitudes toward science, STEAM attitudes, and career choices]* [Master's thesis]. Marmara University.

Helvacı, İ. (2019). *Görsel sanatlar eğitiminde STEAM temelli yaklaşımın etkisi [Effects of a STEAM- based approach in visual arts education]* [Doctoral dissertation]. Gazi University.

Institute for Arts Integration and STEAM. (2019). *What is STEAM education? The definitive guide for K–12 schools.* https://artsintegration.com/what-is-steam-education-in-k-12-schools/

Kolsuz, S. (2018). *Sosyo-bilimsel konuların işlenmesinde steam uygulamaları [STEAM applications in the processing of socioscientific issues]* [Master's thesis]. Afyon Kocatepe University.

Maeda, J. (2013). STEM + art = STEAM. *The STEAM Journal, 1*(1), Article 43. https://doi.org/10.5642/steam.201301.34

Martin, L. (2015). The promise of the maker movement for education. *Journal of Pre-College Engineering Education Research, 5*(1), Article 4. https://doi.org/10.7771/2157-9288.1099

Melfi, T. (Director). (2016). *Hidden figures* [Film]. 20th Century Fox.

Mercan, Z. (2019). *Erken steam geleceğe hazırlık programının* çocukların *görsel uzamsal akıl yürütme becerilerine etkisi [Effects of early STEAM future preparation program on children's visual reasoning skills]* [Doctoral dissertation]. Gazi University.

Ministry of Education. (2018). *Görsel sanatlar dersi* öğretim *program [Visual arts curriculum]*. Author.

Ministry of National Education. (2020). *STEAM education in flexible learning zones.* Ministry of National Education–General Directorate of Innovation and Educational Technologies. http://yegitek.meb.gov.tr/meb_iys_dosyalar/2020_10/15171214_STEAM_Education_in_Flexible_Learning_Zones_-_Article_-_EN.pdf

Miyazaki, H. (Director). (2013). *The wind rises* [Film]. Studio Ghibli.

Mobius kaleidocycles: Sensational structures with potential applications (2018, December 17). https://phys.org/pdf464259451.pdf

Özcan, B. (2019a, September 18). *2050 yılında dünya nasıl bir yer olacak? [What will the world be like in 2050?]* [Video]. https://www.sosyal medya.com/watch?v=rt3ZthtLbWQ

Özcan, B. (2019b, August 18). *Böyle bir insan yok! [There is no such person!]* [Video]. https://www.sosyal medya.com/watch?v=N4m_xt4vFwY

Özcan, B. (2019c, December 15). *İlk 8k vlog: Güney Kore 2.bölüm: Dijital bir kent [First 8k VLOG: South Korea Part 2: A Digital City]* [Video]. https://www.sosyal medya.com/watch?v=7HERFTSaBcU

Rolling, J. H., Jr. (2016a) Reinventing the STEAM engine for art + design education. *Art Education, 69*(4), 4–7. https://doi.org/10.1080/00043125.2016.1176848

Rolling, J. H., Jr. (2016b). STEAM locomotion. *Art Education, 69*(6), 4–5. https://doi.org/10.1080/00043125.2016.1224811

Schönke, J., & Fried, E. (2019). Single degree of freedom everting ring linkages with nonorientable topology. *Proceedings of the National Academy of Sciences, 116*(1), 90–95. https://doi.org/10.1073/pnas.1809796115

Sönmez, S. ve Şahinkayası, Y. (2021). Maker öğretmenlerin maker hareketi ve robotik kodlama faaliyetlerine ilişkin görüşleri [Maker teachers' opinions on maker movement and robotic coding activities]. *Mustafa Kemal* Üniversitesi *Eğitim Fakültesi Dergisi*, *5*(7), 277–296.

Uştu, H. (2019). İlkokul *düzeyinde bütünleşik stem / steam etkinliklerinin uygulanması: sınıf* öğretmenleriyle *bir eylem araştırması* [*Application of integrated STEM/STEAM activities at the primary level: Classroom action research with teachers*] [Doctoral dissertation]. Necmettin Erbakan University.

What is Arduino? (2018, February 5). https://www.arduino.cc/en/Guide/Introduction

Yakman, G. (2010, August 7). *What is the point of STE@M?—A brief overview.* https://www.academia.edu/8113832/What_is_the_Point_of_STEAM_A_Brief_Overview_of_STEAM_Education

CHAPTER 16

Teaching STEAM Education in Hong Kong

So Lan Wong

Abstract

Science, technology, engineering, and mathematics (STEM) education in Hong Kong was proposed by Carrie Lam, the Chief Executive of Hong Kong in 2015 and later optimized in 2022 as science, technology, engineering, arts, and mathematics (STEAM) education. The goal of STEAM education is to cultivate students' solid STEM knowledge to support Hong Kong's rapid economic and technological development to meet contemporary world challenges; however, the role and educational purpose of the arts (i.e., the "A") in STEAM education have not been stated clearly. I share three teaching examples and discuss how to integrate the arts into STEM education and cultivate students' creativity through hands-on activities. To implement STEAM education, teachers could begin with the subjects they teach and then explore ways to integrate the other four areas to create STEAM education.

Keywords

STEAM education – hands-on activities – inquiry learning

1 Introduction

Following the vigorous development of information technology (IT) in the past two decades, science, technology, engineering, and mathematics (STEM) and science, technology, engineering, arts, and mathematics (STEAM) education have gradually assumed their places as global education models. To align with this global trend, policy addresses by the Hong Kong Special Administrative Region (HKSAR; 2015, 2016) announced the implementation of STEM education in Hong Kong. The HKSAR government provided one-time grants of HK$100,000 (approximately US$12,900) to each primary school and HK$200,000 to each secondary school in 2016 and 2017, respectively, to support school-based STEM education (Tang, 2017).

 | DOI:10.1163/9789004714748_017

At the end of 2015, the Curriculum Development Council (CDC), an executive department under the Hong Kong Education Bureau (EDB) responsible for formulating and implementing the Hong Kong education curriculum, issued a planning document called *Promotion of STEM Education: Unleashing Potential in Innovation*. In the document, the EDB proposed strategies for promoting STEM education and recommended that Hong Kong primary and secondary schools develop STEM education in their school-based curricula over the next 10 years.

Eight years have now passed since the release of these policy documents and resources designed to promote STEM education. As a visual arts teacher educator, I asked what foci of STEM education were promoted by the Hong Kong EDB. Why do we advocate for the inclusion of the arts (i.e., the "A") in STEM education, and how well has STEAM education been implemented in Hong Kong? In the following, I briefly introduce STEM education in Hong Kong and the STEAM curricula that I teach in my classrooms.

2 Hong Kong STEM Education: Purpose and Approaches

According to the STEM education documents issued by the Hong Kong EDB (Curriculum Development Council [CDC], 2015, 2016), the promotion of STEM education was designed to maintain Hong Kong's international competitiveness and fulfill its economic, scientific, and technological development needs; therefore, Hong Kong students must be equipped with the skills required for the 21st century, such as the ability to face global changes and challenges. In response to HKSAR's 2015 policy address, the Hong Kong EDB immediately conducted a school survey from November 2015 to January 2016, soliciting opinions from STEM teachers (CDC, 2016). The Hong Kong EDB clearly stated that it would renew the STEM curricula and learning activities in Hong Kong. A 2017 update of the Ongoing Renewal of the School Curriculum webpage by the EDB included a review of most of the Key Learning Area (KLA) Curriculum Guides. Concerning arts education, the CDC (2017a) also added a STEM education title in Section 2.2.6 of Chapter 2: Curriculum Framework. In *Arts Education: Key Learning Area Curriculum Guide (Primary 1–Secondary 6)*, the EDB clearly stated that "in the Hong Kong curriculum context, STEM education is promoted through the KLAs of Science education, Technology education and Mathematics education" (CDC, 2017a, p. 17). Obviously, arts education should also assist STEM-related disciplines in promoting STEM education.

In summary, the aims of STEM education in Hong Kong are to enhance students' interest in STEM and to develop a solid knowledge base in the respective subject areas so that they can study further and pursue careers in response to the rapid economic, scientific, and technological developments, changes, and challenges in the contemporary world. The EDB recommended that all KLAs, such as the arts education KLA, "contribute to the promotion of STEM education through incorporating elements of STEM into arts learning activities" (CDC, 2017a, p. 17).

The findings from the EDB's survey report (CDC, 2016) in the STEM education fields listed its development directions in Hong Kong according to the following two approaches described in *Promotion of STEM education: Unleashing potential in innovation* (CDC, 2015, p. 6).

– Approach 1: Learning activities based on KLA topics for students to integrate relevant learning elements from other KLAs (see Figure 16.1).
– Approach 2: Projects for students to integrate relevant learning elements from different KLAs (see Figure 16.2).

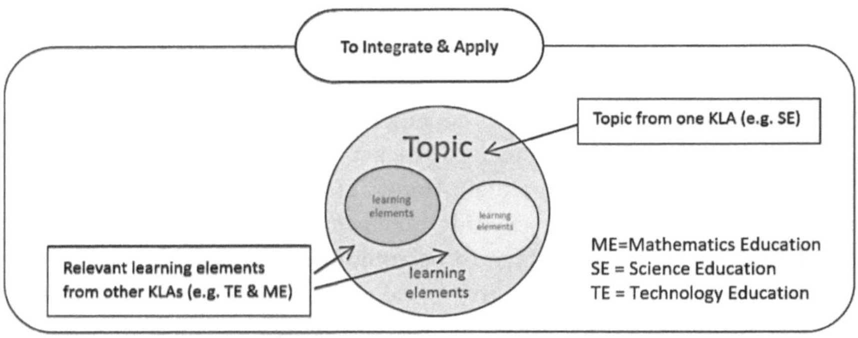

FIGURE 16.1 Learning activities based on KLA topics

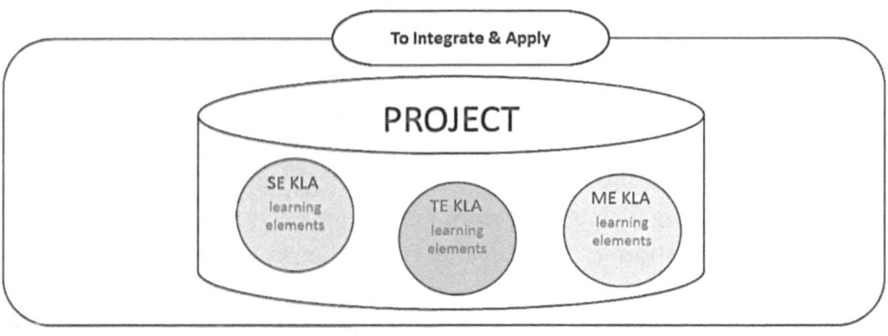

FIGURE 16.2 Learning activities based on projects

The EDB collected 2,584 questionnaires from STEM educators (a response rate of 81.9%). Results showed that the respondents had a high degree of support for the implementation of STEM education. The percentage of teachers who strongly agreed with Approach 1 was higher than that for Approach 2, 90.5% and 81%, respectively (CDC, 2016). I can relate to this result. Taking the two STEAM teaching projects I initiated in the past 3 years as examples, I found designing, implementing, and publishing A-focused STEAM education (Wong, 2021, 2023) easier than interdisciplinary STEAM education (Wong, 2022) because in the former I can insert other STEM-related learning activities into the "A" teaching area with which I am familiar; therefore, I can adjust my arts-focused STEAM instructional design according to my own ability. Regarding the latter, STEAM education based on project learning poses many difficulties that cannot be resolved with only single-disciplinary knowledge. In implementing this interdisciplinary STEAM project, I sought advice about STEM education from my colleagues.

3 The Power of Including "A" in STEM Education

Recent scholars have increasingly explored the differences between STEM and STEAM education (Perignat & Katz-Buonincontro, 2017; Quigley & Herro, 2019; Sousa & Pilecki, 2013, 2018). After tracing the advent of STEM education and the reasons for changing to STEAM education, Bequette and Bequette (2012) pointed out that STEM-related disciplines were focused on arguing from evidence, problem solving, and reconciling multiple solutions; but arts disciplines emphasized aesthetic and design thinking processes. In addition, making art entails a constant search for possible creative and imaginative visual solutions to problems in the design process, and arts disciplines offer students opportunities to express their ideas, feelings, and personal meaning.

Like Quigley and Herro (2019) I believe that moving from STEM to STEAM education can present students with powerful and authentic learning opportunities. Providing occasions to express oneself using visual languages and enabling the persistent pursuit of aesthetic improvement are the strengths of the visual arts discipline. The interdisciplinary approach of STEM education in combination with the arts not only helps demonstrate the students' creative thinking process but also cultivates their willingness to modify their creative products, just like the persistence of artists and scientists in achieving perfection.

Many countries have promoted STEAM education in recent years, but I particularly appreciate the Korean government's advocacy of STEAM education because its policy clearly describes the power of including "A" in STEAM education. The Korean government found that Korean students lacked interest, enjoyment, and confidence in science and mathematics and that they were less motivated and valued science and mathematics less than the average level set by the Programme for International Student Assessment and Trends in International Mathematics and Science Study. When Korean government leaders recognized the value of the characteristics of arts education, they decided to add the "A" to STEM education in 2011, officially launching their STEAM education policy, which clearly stated the necessity of cultivating students' creative and artistic talents. STEAM education in South Korea is designed to cultivate geniuses like Leonardo da Vinci and Steve Jobs among new generations of Korean students (Jon & Chung, 2015). The Korean government used these two well-known innovators as examples to facilitate people's grasp of all that the transition from STEM to STEAM education could bring to Korea.

Leonardo da Vinci and Steve Jobs were considered to have similar personalities and were deeply appreciated as creative geniuses during their lifetimes. These curious, imaginative, creative perfectionists focused their widely ranging knowledge on achieving breakthroughs (Burrows, 2004; Grant, 2018; Mason, 1993). As an engineer and artist, da Vinci was surrounded by mathematicians, architects, and artists in the late 1400s. Similarly, the 1970s San Francisco Bay Area environment gave Jobs the opportunity to start his Apple computer business in pursuit of perfection in craftsmanship, simplicity, and efficiency (Isaacson, 2011, 2017). I appreciate the Korean government's clear goal of promoting STEAM education, providing good school environments, and cultivating the creativity of the next generation of Korean students.

4 Teacher Training and Sample Activities for STEAM Education in Hong Kong

The Education University of Hong Kong (EdUHK) is its main teacher education institution, founded to train primary and secondary school teachers. The primary mission of the EdUHK (2022) is to lead educational innovation and to promote and support the strategic development of teaching, teacher education, and disciplines complementary to education by preparing outstanding and morally responsible educators. Based on the role of the EdUHK and in response to the development trend of international education and curriculum reform in Hong Kong, STEM-related departments have gradually added STEM and STEAM education courses to cultivate forward-looking educators.

Since 2015 I have taught a course at EdUHK called Nurturing Creativity Through Visual Arts Activities, incorporating STEAM elements since 2017. This elective undergraduate course is part of the Creative Minor Programme for various disciplines at EdUHK. The course aims to stimulate preservice teachers in various majors to master the teaching methods of "learning by doing" and "inquiry learning" through visual arts and STEAM activities. In this course, I invite preservice teachers to experience creative learning and inquiry through hands-on activities, such as making light-up cards and playing music on drawings to explore closed circuits, allowing preservice teachers to think about how to use their learning experience in STEAM-related activities to prepare for their future work as teachers at primary and secondary schools.

4.2 *Making Light-Up Cards*

Preservice teachers typically prefer to adopt teacher-led teaching methods; therefore, I designed teaching activities focused on guiding them to experience how to transform what they want to "teach" into visual arts or STEAM-related inquiry learning activities. After I distribute the required materials, including light-emitting diodes (LED), lithium batteries, copper foil tape, transparent tape, A5 drawing paper, and colored markers to preservice teachers, the inquiry learning activities proceed as follows:

Activity 1: Master a simple closed circuit
- Inquiry and problem solving: I asked preservice teachers to use a lithium battery to make an LED bulb glow and explained how they could connect the battery to the bulb to do so.
- Purpose: Let preservice teachers observe and discover that (1) the LED bulb has long and short legs, and the lithium battery has positive and negative poles; (2) when the long leg is connected to the positive pole and the short leg is connected to the negative pole, the LED bulb lights up; and (3) closed circuits have applications in daily life.

Activity 2: Master the series circuit
- Inquiry and problem solving: I asked preservice teachers to make two LED bulbs glow in series using 1–2 lithium batteries and explained how to connect the LED bulbs to do so.
- Purpose: Let preservice teachers discover that (1) the two LED bulbs should connect the long leg to the short leg; (2) the long leg is connected to the positive pole of the lithium battery, and the short leg is connected to the negative pole; and (3) the reason the LED bulbs did not light up is that more electrical energy is needed when the number of LED bulbs increases.

Activity 3: Master the parallel circuit

- Inquiry and problem solving: I asked preservice teachers to use 1–2 lithium batteries to make two parallel LED bulbs glow and explained how to connect the batteries to make this happen.
- Purpose: Let preservice teachers observe the sample and discover that (1) two LED bulbs should have their long and short legs side by side in the same direction; and (2) the long legs of the parallel LED bulbs are connected to the positive pole of the lithium battery, and the short legs are connected to the negative pole.

Based on the recommendations of the *General Studies Curriculum Guide for Primary Schools* (CDC, 2017b) in Hong Kong, I designed these three warm-up activities to demonstrate to preservice teachers how to integrate their "S" (science, i.e., general studies in Hong Kong) and "A" (the arts, i.e., visual arts learning activities) curricula. I purposefully included the general studies teaching content to show how to conduct STEAM teaching in this subject-based era. After the preservice teachers have mastered these three closed-circuit concepts—from simple to complex tasks—they applied the closed-circuit concept to create light-up greeting cards. Unlike in general studies classes, in which preservice teachers design worksheets to guide their students in exploring concepts through prediction and observation (CDC, 2017b), I used visual learning approaches by asking the preservice teachers to upload photos of their findings to the online bulletin board Padlet.

Greeting cards are commonly used in visual arts education, yet in the era of subject-based education, few visual arts teachers would guide their students to use closed circuits in greeting cards. Likewise, general studies teachers would not ask their students to use closed-circuit principles in a greeting card design after completing worksheets. In the era of STEAM education, however, I agree that teachers must expand their knowledge beyond a single area of expertise to provide space for their students to think creatively. After exploring three closed-circuit concepts through hands-on activities, preservice teachers must choose 1–2 closed-circuit methods to light up their own greeting card designs. They also need to draw the correct circuit route on the inner page of the greeting card according to the number and position of the LED bulbs placed (see Figure 16.3).

In the activities noted above, I found that preservice teachers who were not general studies majors needed more time to successfully light the LED lights on a greeting card. I was amazed that even preservice general studies teachers spent some time thinking about the connection between closed-circuit paths and the spots on greeting cards that needed to be lit. Obviously, the "A"

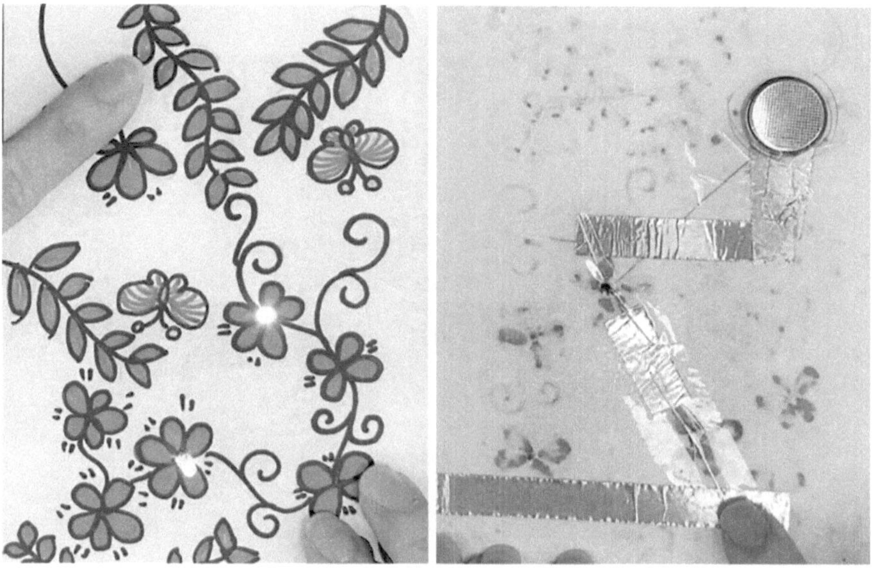

FIGURE 16.3 Front (L) and inner (R) pages of a light-up greeting card

learning activities lend STEAM teaching more space for creative thinking and problem-solving opportunities.

4.2 *Playing Music on Drawings*

After the preservice teachers mastered the closed-circuit concept and created their light-up greeting cards, I provided another STEAM activity that applied the closed-circuit concept further, that is, playing music on drawings using a Makey Makey® STEM Kit, an electronic device that sends commands to a computer in the form of a closed circuit via alligator clips and a USB cable. One of the features of the Makey Makey® Kit emulates a musical keyboard that allows its players to create music. When preservice teachers connect the alligator clips to the conductors correctly, such as pencil drawing, human body, vegetables, and fruits, they can play music using the Makey Makey® Kit.

The materials required for this hands-on activity include a laptop and a Makey Makey® Kit for each group, A3 drawing paper, 6B–8B black lead drawing pencils, and copper foil tape. This STEAM activity opened with individual drawings as follows.

Activity 1: Interweaving lines and imagining images

- Inquiry and problem solving: I instructed the preservice teachers to draw interwoven lines freely and imagine the images that appear. The preservice

FIGURE 16.4 Students' drawings and exploration of various methods to use the Makey Makey® Kit and closed circuits

teachers then formed groups and assembled their images to create a story. They were asked to play a piece of music using their drawings while telling the created story.

- Purpose: Let preservice teachers experience the creative practice of free drawing, allowing them to share and experience "free association" involving forming spontaneous and random thoughts, memories, and feelings inspired by the drawing (BetterHelp, 2023).
- Steps and requirements (see Figure 16.4)
  1. Freely draw organic and geometric lines on A3 drawing paper.
  2. Each line must extend to the boundary of the A3 drawing paper.
  3. Turn the drawing paper to observe and free associate about the images seen in the interwoven lines.
  4. Thicken the outline of the image using 6B–8B pencils or copper foil tape to create a stable conductive field.
  5. Conceive a story from the images created by the group members using forced association (The Daniel J. Hilferty Center for Innovation, 2023), which stimulates creative thinking.

Activity 2: Using closed circuits to make a musical keyboard

- Inquiry and problem solving: After understanding the basic connection methods and functions of the Makey Makey® STEM Kit, I instructed each group to explore how to connect it to their drawings to make sound.

- Purpose: Let preservice teachers connect multiple drawings using the Makey Makey® Kit through trial and error to form a closed circuit that allows Makey Makey® Kit to be used to play music on the drawings.
- Requirements (see Figure 16.4)
  1. Share closed-circuit methods for making sounds using the Makey Makey® Kit on the drawings.
  2. Show the drawings and use the Makey Makey® Kit to play background music and tell a story.

Teaching STEAM using the Makey Makey® STEM Kit may be common in the United States; however, it is rarely used in Hong Kong. I found that most of the preservice teachers could play the piano, but it was a challenge for them to use alligator clips to connect several drawings using 6B–8B pencils or copper foil tape to complete a closed circuit and play a piece of music or tell a story. After trial and error, they ultimately succeeded in using various closed circuits to make sounds using their drawings. They were excited to be able to play a piece of music using the Makey Makey® Kit on their drawing (see Figure 16.5).

My colleagues who teach music consider the Makey Makey® STEM Kit an interesting performance tool. Preservice general studies teachers have noted that primary or secondary school students can use it to distinguish electrical conductors; however, I have spent some time thinking about how to use the

FIGURE 16.5 Using the Makey Makey® Kit and drawings to play a song

Makey Makey® Kit meaningfully in visual arts activities instead of introducing it as a new learning tool. Most preservice teachers are not visual arts majors; therefore, this kind of individual free association drawing—in combination with the groups' forced association activities—can not only relieve the pressure to draw well but also help them to master these two creative theories through visual arts activities. According to feedback from preservice teachers, this activity can inspire them to think about the significance of using multiple STEAM tools in their teaching.

4.3 *Making Tessellation Art*

Incorporating STEAM elements into my Nurturing Creativity Through Visual Arts Activities course allowed me to improve my science knowledge quickly. In the case of the following example, I also learned about the relationship between mathematics and visual arts activities.

When exploring STEAM education topics, the preservice teachers were required to showcase a hands-on instructional design that integrated visual arts or STEAM activities for primary or secondary school students according to their majors. They were asked to use 30-minute microteaching practices to demonstrate how to apply relevant knowledge to daily life through specific examples.

One of the topics was M. C. Escher's method for designing continuous seamless tessellations. These tessellation patterns, such as tessellation tiles, are not difficult to find in daily life; however, few visual arts students have learned about the relationship between M. C. Escher's work and mathematics. Thus, I encouraged a group of preservice mathematics and visual arts teachers to showcase their tessellation designs, which allowed me to learn the tessellation principles from a mathematical perspective.

First, the group showcasing their Escher-inspired designs explained that their teaching targets were 11-year-old fifth graders. Prior knowledge required for learning tessellation designs was mastery of the concepts of geometric shapes and the sums of interior angles. The showcase group members distributed a set of squares (i.e., each angle is 90 degrees), a set of equilateral triangles (i.e., each angle is 60 degrees), a set of regular pentagons (i.e., each angle is 108 degrees), and a set of trapezoids (i.e., the sum of their interior angles is 360 degrees) to each fifth-grade group. The showcase group asked the fifth graders to arrange the geometric shapes into continuous patterns as quickly as possible. In this hands-on activity, it was easy for the fifth graders to discover that unlike the other geometric shapes, the pentagons could not be arranged in a continuous pattern without gaps.

The showcase group asked the fifth graders to discuss the following question: Why can't the pentagons be arranged without gaps? The preservice teachers reminded their students to observe and calculate the relationship between the angle of a regular polygon and 360 degrees. Another hint given to the students was to indicate the degrees of each merged angle, where the sum of their angles at the merged point would result in a continuous pattern without gaps. After some discussion the fifth graders answered:

- "Each angle of a regular polygon should be a factor of 360."
- "The sum of interior angles should be 360 degrees or a factor of 360."
- "The sum of the angles at each merge point should be 360 degrees."

This showcase group summarized the fifth graders' findings and used a hexagon as an example of a geometric shape satisfying the above conditions; no matter how it was cut and reassembled into any new shape (see center of Figure 16.6), the new shape could still be seamlessly merged into a tessellation pattern. The showcase group finally let the 5th grade students verify the creative characteristics of tessellation patterns through practical hands-on activities.

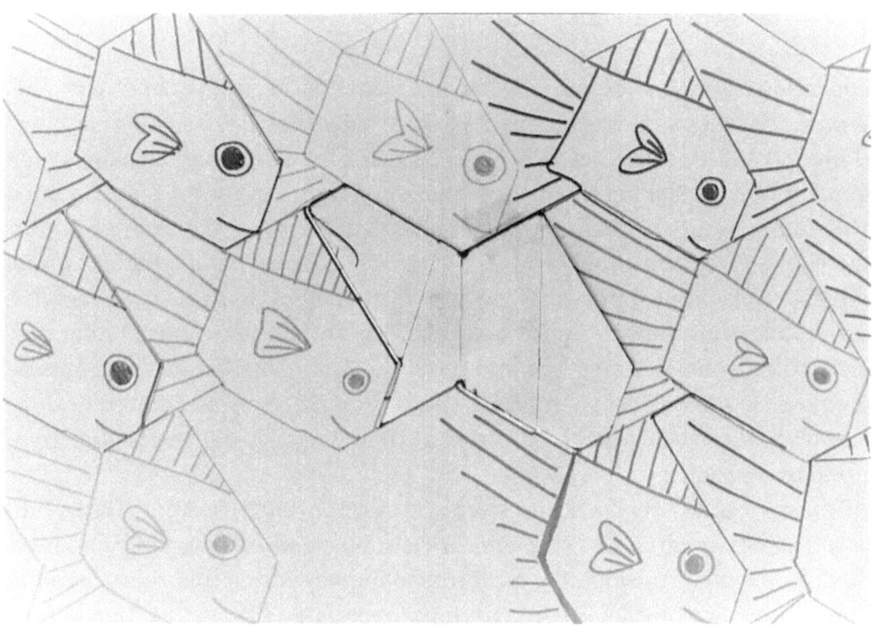

FIGURE 16.6 Using reassembled hexagons to create a new tessellation pattern

Incorporating STEAM education elements into my teaching of the Nurturing Creativity Through Visual Arts Activities course at EdUHK made me rethink the purpose and meaning of learning activities. Taking learning about factors in primary school mathematics as an example, I initially did not understand why I needed to learn factors and how they could be related to my daily life. If my primary school teacher had explained to me that learning about factors could help us to create tessellation patterns like Escher's designs, my interest in learning mathematics would definitely have been aroused, and I may have further explored mathematics subjects.

Another reflection from teaching STEAM education is that single discipline-based knowledge can allow us to master only one-sided concepts or skills. In the case of learning about tessellations, I studied M. C. Escher's tessellation work in secondary school, where my visual arts teacher told us only that Escher's creations were closely related to mathematics, but she did not point out any mathematic techniques that could be applied to Escher's art. Through the showcases by preservice teachers, I obtained STEAM knowledge behind tessellation and simultaneously understood M. C. Escher's art in a rich and meaningful way.

5 Reflections and Conclusion

Educational scholars of various eras have pointed out that school curricula should be integrated and connected with daily life (Bruner, 1960; Darling-Hammond, et al., 2020; Dewey, 1916; Kolb, 1984), but schools and teachers tend to resist change. Taking the comprehensive curriculum reform implemented after Hong Kong returned to China more than 20 years ago as an example, the Hong Kong EDB proposed integrating the existing visual arts and music subjects and adding drama and dance to form a comprehensive arts education curriculum; but visual arts and music teachers at that time reported that they did not know how to integrate and teach the arts-related knowledge outside their own majors. Consequently, the EDB revised its recommendation to allow schools to keep the two subjects—visual arts and music—separate after opposition from teachers.

The current STEM education reform, however, is different. After HKSAR proposed the integration of STEM education subjects in its 2015 policy address, the EDB devoted considerable resources to implementing this policy. In addition to providing development funds for school-based STEM education at each primary and secondary school, the EDB organized training sessions to enable teachers to teach STEM subjects in their schools.

At the end of August 2022, Hong Kong newspapers reported that the HKSAR government planned to optimize STEM education by adding artistic elements and that every Hong Kong primary school could hire a full-time STEAM coordinating teacher with a high salary to promote school-based STEAM education (Rong, 2022). Unlike the South Korean government, which has clearly stated that it wants to nurture future creatives of the caliber of Leonardo da Vinci and Steve Jobs, the EDB's policy document does not clearly state the reason for the revision from STEM to STEAM education. In addition, it does not state its optimization goals. This shift from STEM to STEAM education was quietly presented in a footnote on the first page of the *Education Bureau Circular Memorandum* issued by the EDB (2022a, p. 1): "To act in concert with the policy initiative of the new term of government, the EDB has officially renamed STEM education as STEAM education."

The EDB did not propose a clear direction for development before the Chief Executive delivered the HKSAR's 2022 policy address in October. In Paragraph 117 of that document, titled "STEAM Education," the HKSAR proposed the following three policies to support Hong Kong's promotion of innovation and technology (I&T) in education:

- Promote learning "for all"—More I&T learning elements will be incorporated into the education curriculum, with the aim to have at least 75% of publicly funded schools implementing enriched coding education subjects at the upper primary level and introducing elements of an I&T curriculum, such as artificial intelligence, into the junior secondary curriculum by the 2024–2025 school year.
- Strengthen leadership and coordination—Starting from the current school year, all publicly funded primary and secondary schools are required to designate coordinators to holistically plan STEAM education within and beyond the classroom, and starting from the next school year, to organize or arrange students' participation in quality STEAM activities every year; and
- Enhance professional training—At least 75% of publicly funded primary and secondary schools should organize professional training in STEAM education for their teachers within 2 years.

According to the EDB's (2022b) figures and statistics, 591 primary schools, of which 477 were publicly funded, and 508 secondary schools, of which 451 were publicly funded, operated in Hong Kong in 2021–2022. The implication is that 928 primary and secondary schools had to appoint STEAM coordinators for the 2022–2023 school year to plan quality STEAM activities. In addition, about 700 primary and secondary schools were required to arrange for their teachers to participate in STEAM professional training sessions within 2 years and to implement coding and artificial intelligence education in STEAM-related subjects in primary and secondary schools, respectively. Hence, the focus of STEAM education in Hong Kong is consistent with the purpose of STEM education proposed in 2015 and 2016; that is, STEAM education should mainly be used to develop Hong Kong's innovative technological capacity.

Implementing STEAM education in Hong Kong is imperative; however, the focus of optimizing STEAM education in 2022 is still coding and AI education, which are inclined toward technological development. What is the role of "A" in STEAM education? How STEAM education will be understood and implemented in Hong Kong primary and secondary schools and what research is required in the future remains to be seen.

References

Bequette, J. W., & Bequette, M. B. (2012). A place for art and design education in the STEM conversation. *Art Education, 65*(2), 40–47.

BetterHelp. (2023, January 17). *Free association: What is it, and how does it work?* https://www.betterhelp.com/advice/psychologists/free-association-what-is-it-and-how-does-it-work/

Bruner, J. S. (1960). *The process of education.* Harvard University Press.

Burrows, P. (2004, November 1). Steve Jobs: He thinks different. *BusinessWeek Online.* https://web.archive.org/web/20041031014819/http://www.businessweek.com/magazine/content/04_44/b3906025_mz072.htm

Curriculum Development Council. (2015). *Promotion of STEM education: Unleashing potential in innovation.* Education Bureau, Hong Kong Special Administrative Region. https://www.edb.gov.hk/attachment/en/curriculum-development/renewal/Brief%20on%20STEM%20(Overview)_eng_20151105.pdf

Curriculum Development Council. (2016). *Report on promotion of STEM education: Unleashing potential in innovation.* Education Bureau, Hong Kong Special Administrative Region. https://www.edb.gov.hk/attachment/en/curriculum-development/renewal/STEM%20Education%20Report_Eng.pdf

Curriculum Development Council. (2017a). *Arts education: Key learning area curriculum guide (Primary 1 – Secondary 6).* Education Bureau, Hong Kong Special Administrative Region. https://www.edb.gov.hk/attachment/en/curriculum-development/kla/arts-edu/curriculum-docs/AE_KLACG_Eng_2017.pdf

Curriculum Development Council. (2017b). *General studies curriculum guide for primary schools (Primary 1 – Primary 6).* Education Bureau, Hong Kong Special Administrative Region. http://www.edb.gov.hk/attachment/en/curriculum-development/cross-kla-studies/gs-primary/GSCG_2017_Eng.pdf

Darling-Hammond, L., Flook, L., Cook-Harvey, C., Barron, B., & Osher, D. (2020). Implications for educational practice of the science of learning and development. *Applied Developmental Science, 24*(2), 97–140.

Dewey, J. (1916). *Democracy and education: An introduction to the philosophy of education.* The Macmillan Company.

Education Bureau. (2022a). *Education Bureau circular memorandum—Ongoing renewal of the school curriculum: The primary education curriculum guide (pilot version)* No. 154/2022. https://applications.edb.gov.hk/circular/upload/EDBCM/EDBCM22154E.pdf

Education Bureau. (2022b). *Figures and statistics.* https://www.edb.gov.hk/en/about-edb/publications-stat/figures/index.html

Grant, A. (2018, January 17). *From Leonardo da Vinci to Steve Jobs: The benefits of being a misfit.* University of Pennsylvania. https://knowledge.wharton.upenn.edu/article/leonardo-da-vinci-steve-jobs-benefits-misfit/

Hong Kong Special Administrative Region. (2015, January 14). *2015 policy address.* https://www.policyaddress.gov.hk/2015/eng/p150.html

Hong Kong Special Administrative Region. (2016, January 13). *2016 policy address.* https://www.policyaddress.gov.hk/2016/eng/p87.html

Hong Kong Special Administrative Region. (2022, October). *The chief executive's 2022 policy address.* https://www.policyaddress.gov.hk/2022/en/policy.html

Isaacson, W. (2011). *Steve Jobs.* Simon & Schuster.

Isaacson, W. (2017). *Leonardo da Vinci.* Simon & Schuster.

Jon, J. E., & Chung, H. I. (2015). From STEM to STEAM. In B. Freeman, S. Marginson, & R. Tytler (Eds.), *The age of STEM: Educational policy and practice across the world in science, technology, engineering and mathematics* (pp. 33–46). Routledge.

Kolb, D. A. (1984). *Experiential learning: Experience as the source of learning and development.* Prentice-Hall.

Mason, A. (1993). *Famous artists: Leonardo Da Vinci.* Watts Books.

Perignat, E., & Katz-Buonincontro, J. (2018). From STEM to STEAM: Using brain-compatible strategies to integrate the arts. *Arts Education Policy Review, 119*(2), 107–110.

Quigley, C., & Herro, D. (2019). *An educator's guide to STEAM: Engaging students using real-world problems.* Teachers College Press.

Rong, Y. R. (2022, August 29). 教育局*STEM*加入藝術元素准小學彈性聘未受師訓人士任教助推廣 [*Education Bureau adds artistic elements to STEM and approves primary schools to flexibly hire people without teacher training to help promote STEAM*]. https://www.hk01.com/sns/article/809108

Sousa, D. A., & Pilecki, T. (2013). *From STEM to STEAM: Using brain-compatible strategies to integrate the arts.* Corwin.

Sousa, D. A., & Pilecki, T. (2018). *From STEM to STEAM: Brain-compatible strategies and lessons that integrate the arts.* Corwin.

Tang, P. (2017). *STEM education in secondary schools.* Legislative Council of the Hong Kong Special Administrative Region of the People's Republic of China. https://www.legco.gov.hk/research-publications/english/essentials-1617ise13-stem-education-in-secondary-schools.htm

The Daniel J. Hilferty Center for Innovation (2023). *Forced association.* Independence Blue Cross. https://innovation.ibx.com/tools/forced-association/

The Education University of Hong Kong. (2022). *Vision and mission.* Author. https://www.eduhk.hk/en/about/vision-and-mission

Wong, S. L. (Ed.). (2021). 以*A*為焦點的*STEAM*教學 [*An art-focused STEAM education*]. EdUHK.

Wong, S. L. (Ed.). (2022). 綜合的*STEAM*教學 [*An interdisciplinary STEAM teaching*]. EdUHK.

Wong, S. L. (Ed.). (2023). 視覺藝術與*STEAM*教學：從面授轉向線上教學 [*Visual arts and STEAM teaching—Transferring from face-to-face to online teaching*]. EdUHK.

CHAPTER 17

Developing Visual Art Workshops for STEM Teachers

Kevin Hsieh and Michael White

Abstract

As the visual art professional development (PD) providers, we share our experience developing visual art workshops for 20 STEM teachers seeking to integrate visual art programming into their curriculum. Georgia State University (GSU) collaborated with Drew Charter School (a STEAM-certified K–12 charter school in Atlanta), the Georgia Institute of Technology, and Decatur Makers (a local makerspace initiative) on a grant-funded project to develop STEAM PD workshops for 20 K–8 STEM teachers. These workshops centered on the content and pedagogy for integrating visual art into STEM or other subjects. The workshops aimed to equip non-art teachers with simple strategies like art appreciation, art vocabularies, object-based teaching and learning, and hands-on art activities. We found both skill-based and concept-based components essential and helpful for workshop participants because they can quickly adopt the newly acquired skills and knowledge into their teaching practices. Most importantly, teachers' positive attitudes toward art integration are key for schools that promote STEAM curriculum. At the end of the chapter, we share our recommendations for teacher PD providers and K–12 schools that implement transdisciplinary STEAM curricula.

Keywords

visual art workshop – teacher professional development – art integration – STEAM workshop

1 Introduction

In the United States, STEM–STEAM education has garnered attention from K–12 schools and educational agencies like museums and community organizations. STEAM education in Georgia, for instance, has received greater attention with the number of STEM–STEAM certified schools in Georgia growing

 | DOI:10.1163/9789004714748_018

1,033% from 2012 to 2018 (Georgia Department of Education, n.d.a). With an increasing number of K–12 schools adopting STEAM education objectives, helping in-service teachers identify connections between their respective subjects and art is an important task, hence the "A" in STEAM-centric program development. According to the Georgia Department of Education (n.d.b) "arts are instrumental in teaching creative problem solving, innovation, and empathy. The arts provide hands-on avenues for students to engage with curriculum and prepare students for future careers in the growing creative economy." Access to STEAM-focused teacher professional development (PD) that directly addresses content and pedagogy in support of STEAM integration has become an essential task, especially for STEM teachers with no background or training in the arts.

Research studies have shown the benefits of STEAM education for K–12 learners' development of problem solving and design thinking skills (Bequette & Bequette, 2012; Berk, 2016; Hsieh et al., 2019; Liao, 2016), encouraging their creative investigation and becoming meaningful makers (Patton & Knochel, 2017) as well as acquainting them with 21st-century skills (Graham, 2020; Liao et al., 2016). For instance, environmental studies teachers can use visual representations and visualizing science data to guide students to develop solutions for preserving water sources. Researchers have also identified numerous benefits of integrating the arts into STEM coursework, including helping K–12 learners bridge the gap between various disciplines (Herro & Quigley, 2017; Jamil et al., 2018; Liao et al., 2016). For instance, math teachers can use visual diagrams or animations to explain abstract mathematical concepts. English teachers can use visual journals to enhance students' reading and writing skills. The arts are also "seen as more equitable for communities that are typically excluded [and] the inclusion of art, creativity and design skills is considered appealing to diverse populations of learners such as girls and students of color" (Herro et al., 2018, p. 486). According to the annual report of the Drew Charter School (a STEAM-certified K–12 public charter school in Atlanta), students in Grades 3–8 at Drew have higher academic performances on English language arts, mathematics, science, and social studies compared to the State and Atlanta Public Schools averages. Drew School's student diversity is 85.9% minority enrollment, which includes 78.2% Black or African American students (U.S. News, 2023).

2 Project Background

In 2018 the School of Art and Design at Georgia State University (GSU) collaborated with Drew Charter School, the Georgia Institute of Technology (GA Tech), and Decatur Makers (a local nonprofit makerspace) on a grant-funded three-year project titled STEAM Beyond Borders (SBB). The objective of SBB was to develop a broad range of STEAM PD workshops for a cohort of 20 K–8 STEM teachers from four K–12 schools across the state. Twenty teachers with backgrounds in business, computer science, dance, engineering, English language arts, environmental science, mathematics, music, science, Spanish, reading, and robotics participated in various visual art PD workshops offered over three years. The grant period ran from summer 2018 to spring 2021. GA Tech provided workshops focusing on the hard sciences, Decatur Makers offered hands-on workshops emphasizing engineering, and GSU delivered traditional visual art workshops. Primary investigators (PIs) from GSU in charge of the visual art portion of the SBB project included an art education professor (Kevin) and a design professor and architect (Michael). Michael brought department-level leadership and administrative experience while Kevin brought experience with K–12 art teacher education and art learning theories. In addition to being co-PIs for the project, Kevin and Michael were also in charge of developing and evaluating all aspects of the GSU visual art workshops, including instructor recruitment, workshop scheduling, material procurement and distribution, workshop content development, content alignment with K–8 education visual art standards, standardized documentation of all workshop offerings and activities, on-site and remote recording of workshop demonstrations and activities, and postworkshop survey creation and reporting.

As the primary visual art PD providers on this project, we understood that identifying meaningful connections between art and the various academic disciplines was essential for both teachers and learners (Guskey, 2002). We were specifically tasked with developing and offering a total of 54 visual art workshops (approximately 18 workshops per academic year) to the K–8 teacher cohort. Converting these 20 STEM teachers into art teachers was not the goal. The main goal of workshop development was to assist non-art teachers in making connections between visual art and their academic disciplines; furthermore, the workshops provided a space for open discussion among participants on how they might integrate the arts into their teaching practices.

3 Content Development of Visual Art Workshops

During the planning phase of workshop development in spring 2018, understanding the needs of these 20 STEM teachers was critical for us. Through in-depth monthly discussions with the Drew Charter School team, comprising the director of fine arts & STEAM, the STEAM grant facilitator, the grant coordinator, and STEM teachers from within the cohort, a wide range of art and design subjects of interest were identified, including digital animation, city planning, graphic design, illustration, sketchbook creation, object-based learning from art history, 2D design printmaking, design thinking, and textile printing. Subsequently, we recruited School of Art and Design faculty members from those subject areas to lead the visual art PD workshops, sketchbook creation (see Figure 17.1) and 2D design (see Figure 17.2) for instance. One important aspect was to communicate to the GSU workshop faculty the specific non-art backgrounds of the workshop participants because we expected workshop teachers to present instruction at a level easily understood by non-art teachers. GSU offered a slate of 15–18 visual art PD workshops for teachers to choose from and enroll in each year. Each workshop was 90 minutes in length and was scheduled either after regular school hours or on Saturday mornings. The 20 STEM teachers were required to participate and complete at least 32 PD hours per academic year and to complete at least one workshop from each PD provider. Teachers could choose from offerings by any of the PD providers: GSU, GA Tech, or Decatur Makers. For example, a teacher might select one sciences workshop, one maker workshop, and 14 visual art workshops or some

FIGURE 17.1 In the Making Sketchbooks Workshop, 2020, the instructor gave a step-by-step demonstration on how to use recycled cereal boxes and copy paper to make sketchbooks

FIGURE 17.2 Drawing and painting lecturer showed the processes of 2D design and making patterns

other combination based on interest. Teachers were also allowed to exceed the 32 PD hours.

The visual art PD workshops were conducted in person either on the GSU campus or at the high school campus of the Drew Charter School before the 2020 COVID-19 pandemic. GSU recorded and streamed each workshop for those teachers unable to travel to the Drew Charter School in Atlanta. After the midspring 2020 shutdown due to the pandemic, we switched the workshop modules from in-person to fully synchronous online sessions. This unplanned modality shift had significant effect on content development and delivery as compared to initial plans, but teacher assessment and evaluations per the post-workshop surveys continued to yield constructive input from teachers as they, too, shifted modalities in their own classrooms. Throughout the grant period our strategy for developing relevant workshop content and focus was derived from teacher feedback and recommendations collected through annual pre-planning sessions and more immediate postworkshop surveys as part of the SBB project assessment protocol.

After the first year (2019) of in-person, digital-centric, skill-building workshops and workshops addressing art-making fundamentals, PD participants recommended that we modify our offerings to privilege two objectives: the first, the inclusion of more hands-on activities with takeaways (art materials, art-making tools, and actual projects); and the second, the allowance of more time within the workshop for identifying viable connections between art and traditional disciplines, such as mathematics, social studies, music, or sciences. Given this preference for workshops with simple hands-on activities using readily available classroom materials, such as making a sketchbook or designing 2D patterns with simple rulers, compasses, or templates (see Figure 1), and a desire for structured group discussion regarding direct connections between visual art and their respective disciplines, we divided our visual art PD workshop content development into two categories: skill-based and concept-based. We describe both below.

4 Skills-Based or Hands-On Content

In skill-based or hands-on workshops we focused on the introduction of basic art skills, such as drawing simple lines and shapes, designing patterns, mixing colors, calligraphy, or collaging. This aligns with the research of Harvey Seifter and colleagues as outlined more extensively through the National Science Foundation-funded initiative called The Art of Science Learning. They claimed: "Arts-based learning is the instrumental use of artistic skills, processes and experiences as educational tools to foster learning in non-artistic disciplines and domains" (The Art of Science Learning, n.d., para. 1). Incorporating hands-on, imaginative approaches to science education and using many of the methods associated with the creative arts have been shown to attract and retain young people in the fields of science, technology, engineering, and mathematics.

In a Sculptural Hat Workshop taught virtually by an art education professor, teachers were introduced to the historical and cultural significance of hats over time and in various societies. Step-by-step demonstrations on how to use construction paper to build a wearable three-dimensional hat included simple exercises, such as measuring head size and folding, cutting, and constructing a hat. As shown in Figure 17.3, teachers made a variety of sculptural hats using simple construction paper and tape and then decorated them with materials easily found in their homes. The hands-on practice helped PD participants to think or plan what they could do in their future teaching. Participant #12 shared the following in the postworkshop survey: "I plan on using this lesson [sculptural hats] to help my students with paper folding and forms to teach 3D shapes" (personal communication, May 15, 2021). In Puppetry Arts: Making Art

FIGURE 17.3 In the Sculptural Hats Workshop (2021) taught by an art education professor, the participants showed their three-dimensional paper hats

With Found Objects, taught by the same art education professor, Participant #15 stated in the postworkshop survey: "I teach all elementary subjects. I could see myself using this course [puppetry arts] to teach about historical figures [making puppets of them]. I could also see it being used for science—on using recycled materials" (personal communication, May 25, 2021).

In addition to learning hands-on art-making activities, teachers also developed observational and visual analysis skills. Taught by an art historian, the Object-Based Learning Workshop introduced participants to simplified methods of visual analysis and documentation of found objects or artworks. Exercises designed to hone skills of observation in students further showed how visual analysis could be used to formulate expanded research questions. In this workshop the instructor introduced and explained various elements of art and principles of design that might be used to describe an object and then apply those same elements and principles to analyze 3D and 2D art works.

The Object-Based Learning Workshop echoed the work of Carpenter and Sessions (2002), who described using a content-based approach to analyze three-dimensional artworks. Three essential areas appear in their model: technique, context, and concept. They pointed out that "the content of an artwork might explore specifically how the work was created (technique), the influential social and cultural events relevant to that work (context), and the philosophical implications of the content of the work (concept)" (p. 373). Familiarizing these 20 STEM teachers with specific visual art analysis skills—technique, context, and concept—became a fundamental objective of subsequent PD development efforts.

5 Concept-Based Content

In art, concept refers to themes, topics, issues, or concerns, shown in or derived from artworks. Concept is the "big idea" that drives learning tasks. The big idea is a term used by art educators when they are developing curriculum or lessons (Stewart & Walker, 2005). Sickler-Voigt (2020) described big ideas in an art education context as "broad topics that [learners] study in depth to address significant human issues that remain relevant regardless of the times and cultures in which people live" (pp. 6–7). Some examples of big ideas might include peace, displacement, environmental issues, transportation, time, gamification, fashion, tradition, accessibility, play, or innovation. Big ideas not only assist teachers in developing meaningful lessons but also guide students "in grasping the meaning and reasoning behind the big ideas that drive learning tasks" (Sickler-Voigt, 2020, p. 7). Taking this big ideas concept as a departure point, participants identified one to two big ideas in the planning session as themes around which workshops were developed. One to two big ideas were identified each semester, yielding three to six big ideas over the academic year.

For example, one series of workshops focused on environmental, community, and social justice issues (e.g., Black Lives Matter movement). For this objective, we expanded our roster of instructors beyond our immediate university community and engaged local practicing Black artists, architects, and designers to offer workshops built around examples of their current work, focusing on the concept or big idea of social justice. Workshops offered in fall 2020 included the following:

- Black Fungibility and Collage, taught by a local Atlanta artist whose work explores ideas of alienation, identity, and Black embodiment;
- Art as Agent of Community Identity and Change, taught by another artist whose narrative works explore historical and contemporary social concerns and issues from the female perspective;
- Art Activism: Ocean Conservation and Plastics, taught by a member of the Georgia State University faculty and artist whose sculptural found-plastic installations both document and highlight the impact of plastic pollution on delicate ecosystems; and
- Urban Planning and Architecture, taught by an architectural scholar and historian who showed the impact of urban planning policies over time in Atlanta—with both planned and unplanned ramifications.

In each of these workshops instructors described, analyzed, interpreted, and discussed the ideas and concepts behind their works and how they visualized and communicated their abstract ideas and concepts during the creative process. In the postworkshop survey one computer science teacher shared the following after attending the Urban Planning and Architecture Workshop: "I would collaborate with the science teacher and have students learn how to test the water and soil in their neighborhoods as they explore how to design their urban gardening plans" (Participant #19, personal communication, November 20, 2020). The same teacher applied the idea of collage to two different disciplines after he participated in the Black Fungibility and Collage Workshop as he wrote, "My tactile and participatory collage may need to utilize music and also electronic components, which would invite collaboration with those subject teachers and/or students."

Implementing both skills-based and concept-based content in art integration PD workshops for STEM teachers gives them powerful pedagogical tools to connect art to their subject areas. More importantly, it also helps STEM teachers see the value of visual art in STEAM education; however, a follow-up study on how and in what way STEM teachers teach their collaborative STEAM lessons in school might be needed.

6 Essential Approach Learned: Art Educator as Liaison

An important strategy implemented in our PD workshop development was to have actual art educators involved during the planning, development, execution, and assessment phases of the PD project. We quickly recognized and learned that beyond basic art-making content, the pedagogical expertise brought to the table by art educators was essential. Before the Black Fungibility and Collage workshop, for instance, we held virtual meetings with the workshop instructor to discuss the various approaches she might use to demonstrate the collage process and explain how artists use the concept of fungibility in creating collage works. During this meeting, Kevin, an art education professor, asked specific questions to elicit clarification, resulting in the artist–instructor incorporating more concrete examples into the workshop to explain abstract concepts. In this and other sessions with instructors, Kevin also discussed the Georgia K–8 visual art standards and encouraged workshop instructors to allow 15 minutes or more for participants to share their thoughts on possible incorporation ideas before dismissing the workshop.

Supported by the guidance from Kevin in the development of visual art PD workshops, workshop instructors were relieved from feeling they needed to be both subject matter experts in art *and* experts in K–8 education. Second, Kevin's guidance set up a partnership with STEM teachers to bring possible in-class content applications to the table, such as the Urban Planning and Architecture Workshop and the Art Activism: Ocean Conservation and Plastics. If we did not have an art education professor like Kevin as a liaison, we would have been challenged to find art instructors willing to teach the workshops, and we might not have been able to break down the visual art content knowledge and possible applications of art for our STEM teachers. Our process relied on a successful implementation of an art education faculty as an intermediary liaison.

Once workshop topics and instructors were identified, Kevin opened a dialogue with each workshop instructor and then developed a summary workshop catalog entry that aligned state visual arts standards with the workshop content and description. More specifically, Kevin learned about the content from workshop instructors and then discussed potential or optional pedagogies that could be adopted to break down the content knowledge in ways that K–8 STEM teachers could both understand and apply in their classrooms. The resulting workshop guidelines outlined and clarified for both the art instructors and the STEM teachers which State Visual Art Standards aligned with workshop content. The guidelines became a key recruitment tool for attracting both art faculty members as instructors and STEM teachers as students to specific workshops. The guidelines demystified unfamiliar K–8 pedagogical objectives for the former and provided a framework for art content integration for the latter. Kevin also shared guiding questions for postworkshop discussions among the participants. Sample questions included the following: How would you implement what you just learned in your subject area? What were the principles of design or elements of art used in this session? How might you see this exercise supplementing content understanding across disciplines? Because more than 90% of the visual art workshop instructors were professional studio artists from GSU with limited K–8 pedagogical training, having an art educator as a liaison to bridge the pedagogical gap became critically important to faculty recruitment.

7 STEM Teachers' Attitudes toward Collaboration with K–8 Art Teachers

"Collaborating and sharing are often valued by participants [teachers] and allow them to establish a greater support network" (Parson et al., 2019, p. 34). From the postworkshop survey we learned that STEM teachers were inspired by the workshop content and were eager to reach out and initiate conversations with visual art teachers at their institutions for collaborative lesson development. We not only saw STEM teachers making connections to visual art content, but we also saw evidence that they valued visual art in K–8 classrooms, an essential part of art advocacy and a core tenet of STEAM education.

One physical science teacher wrote after attending the Black Fungibility and Collage Workshop, "I would collaborate with the visual arts and the orchestra teacher. I have collaborated in the past with the Visual Arts teacher to create plaster cast masks in a social– emotional perception project with my physical science students" (Participant #7, postworkshop survey, November 20, 2020). A language teacher wrote after participating in the Chinese Calligraphy Workshop, "I can work with the dance teacher to create movement interpretation of our words" (Participant #16, postworkshop survey, November 20, 2020). Participant # 11 shared her thoughts on interdisciplinary collaboration after attending the Logo and Branding Workshop: "I could collaborate with the PE teacher to make tableau versions (physical) of the logos after they are created. I could collaborate with the art teacher on this as well to make the logos more interesting" (postworkshop survey, May 20, 2021).

8 Conclusions and Suggestions

In addition to the regular monthly meeting with Drew's team for feedback and suggestions, the mandatory postworkshop survey as an assessment and evaluation tool provided us with both critical and important data for future workshop content development, including first, the expansion of specific workshop content for further explorations (e.g., we created several follow-up workshops, such as Chinese Calligraphy Parts II and III Workshops, for more in-depth learning of specific skills on operating Chinese brushes and ink value) and second, connecting workshop content to contemporary and current sociocultural issues (e.g., the Art as Agent of Community Identity and Change was developed as an exploration of social justice issues in response to this request).

STEAM education emphasizes interdisciplinary teacher–teacher, teacher–student, and student–student collaborations (Guskey, 2002; Liao, 2016). Although STEAM education is an approach developed over the past decade, the idea of transdisciplinary connections (Liao, 2016) and collaborations is not new. With visual art PD opportunities for non-art teachers or STEM teachers, the value of interdisciplinary connection with visual art can be highlighted and reinforced. From a PD provider's perspective in developing visual art PD workshops for teachers with no visual art background, understanding the unique needs of STEM teachers and helping them remove the roadblocks to visual art are essential. We hope that incorporating both skills-based and concept-based visual art content helped non-art teachers to see how artists use their visual works to address ideas and concepts inspired and informed potential connections between art and their various disciplines. In the future STEAM PD should be further investigated to confirm that participants can effectively integrate their ideas into their actual teaching, including pedagogical practices and curriculum assessment designed to provide K–8 students with meaningful and effective learning experiences.

References

Bequette, J. W., & Bequette, M. B. (2012). A place for art and design education in the STEM conversation. *Art Education, 65*(2), 40–47. https://doi.org/10.1080/00043125.2012.11519167

Berk, S. (2016). Designing for the future of education requires design education. *Art Education, 69*(6), 16–20. https://doi.org/10.1080/00043125.2016.1224844

Carpenter, B. S., II, & Sessions, B. (2002). (Re)Shaping visual inquiry of three-dimensional art objects in the elementary school: A content-based approach. In Y. Gaudelius & P. Speirs (Eds.), *Contemporary issues in art education* (pp. 370–383). Prentice Hall.

Georgia Department of Education. (n.d-a). *STEM/STEAM Georgia*. Retrieved August 15, 2020, from https://www.stemgeorgia.org/about-us/

Georgia Department of Education. (n.d-b). *STEM/STEAM Georgia*. Retrieved August 15, 2020, from http://www.stemgeorgia.org/wp-content/uploads/2018/10/STEM_STEAM-in-Georgia.pdf

Graham, M. A. (2020). Deconstructing the bright future of STEAM and design thinking. *Art Education, 73*(3), 6–12. http://dx.doi.org/10.1080/00043125.2020.1717820

Guskey, T. R. (2002). Professional development and teacher change. *Teachers and Teaching: Theory and Practices, 8*(3), 381–391. http://dx.doi.org/10.1080/135406002100000512

Herro, D., & Quigley, C. (2016). Exploring teachers' perceptions of STEAM teaching through professional development: Implications for teacher educators. *Professional Development in Education, 43*(3), 416–438. http://dx.doi.org/10.1080/19415257.2016.1205507

Herro, D., Quigley, C., & Jacques, L. A. (2018). Examining technology integration in middle school STEAM units. *Technology, Pedagogy, and Education, 27*(4), 485–498. https://doi.org/10.1080/1475939X.2018.1514322

Hsieh, K., Glassmeyer, D., & Nguyen, L. (2019). Collaborative teaching and learning: High school STEAM project-based learning and lantern design. *The International Journal of Arts Education, 17*(1), 33–67.

Jamil, F. M., Linder, S. M., & Stegelin, D. A. (2018). Early childhood teacher beliefs about STEAM education after a professional development conference. *Early Childhood Education Journal, 46*(4), 409–417. https://doi.org/10.1007/s10643-017-0875-5

Liao, C. (2016). From interdisciplinary to transdisciplinary: An arts-integrated approach to STEAM education. *Art Education, 69*(6), 44–49. https://doi.org/10.1080/00043125.2016.1224873

Liao, C., Motter, J., & Patton, R. (2016). Tech-savvy girl: Learning 21st-century skills through STEAM digital artmaking. *Art Education, 69*(4), 29–35. http://dx.doi.org/10.1080/00043125.2016.1176492

Parson, S. A., Hutchison, A. C., Hall, L. A., Parsons, A. W., Ives, S. T., & Leggett, A. B. (2019). U.S. teachers' perceptions of online professional development. *Teaching and Teacher Education, 82*(3), 33–42. http://dx.doi.org/10.1016/j.tate.2019.03.006

Patton, R. M., & Knochel, A. D. (2017). Meaningful makers: Stuff, sharing, and connection in STEAM curriculum. *Art Education, 70*(1), 36–43. http://dx.doi.org/10.1080/00043125.2017.1247571

Sickler-Voigt, D. (2020). *Teaching and learning in art education: Cultivating students' potential from pre-K through high school.* Routledge.

Stewart, M. G., & Walker, S. R. (2005). *Rethinking curriculum in art.* Davis Publication.

The Art of Science Learning. (n.d.). *Arts-based learning.* https://www.artofsciencelearning.org/arts-based-learning/

U.S. News. (2023). *Overview of Charles Drew charter school Ja/Sa.* U.S. News Education. https://www.usnews.com/education/k12/georgia/charles-drew-charter-school-ja-sa-200306

Index

3D xvi, 15, 46, 56, 59, 61, 75–77, 79, 89, 113, 114, 153, 161–163, 165, 168, 169, 193, 198–201, 203, 206, 209, 286, 287

accessibility 14, 154, 288
aesthetics 65, 94, 97–100, 102, 104, 105, 145, 149, 197, 203, 227, 230, 231, 233, 234, 236, 243, 256
agency 14, 107, 110, 111, 119, 120, 122
artistic behavior 14, 38, 40, 41, 44, 47, 48
arts integration 2, 3, 24, 108, 109, 250
augmented reality 15, 115, 161, 167, 198, 209, 251, 254
authentic 1, 14, 30, 38–40, 44, 45, 48, 100, 110, 122, 267
authenticity 27, 54, 65, 110

big ideas 40, 45, 288
biology 77, 103, 104, 111, 115, 118, 165, 239

career 12, 46, 87, 105, 110, 111, 118, 154, 196, 266, 282
chemistry 11, 77, 104, 111, 113, 161, 164, 165, 168, 169, 170, 173–176, 239, 255, 259
circuit 193, 194, 220–222, 269–273
coding 13, 43, 77, 85, 206, 212, 218, 219, 226, 249, 251, 253, 256, 259, 277, 278
collaboration xiii, xv, 1, 12, 14, 21, 23, 24, 27, 29, 31, 32, 66, 67, 86, 95, 105, 109–111, 167, 182, 194–197, 201, 203, 289–291
community-based 2, 10, 14, 44, 85, 90, 95, 97
competencies 14, 16, 21–25, 27, 28, 30–32, 79, 105, 143, 197, 211, 249–251, 256
competency 15, 16, 29, 31, 211, 256
computational xvii, 210, 211, 215, 216, 219, 223
critical thinking 21, 22, 24, 25, 27, 29–32, 89, 93, 95, 110, 250

data visualization 126, 127, 129, 131, 133–135, 138, 139, 141–1143, 195
design thinking 2, 6, 10, 13, 14, 21, 26, 27, 29, 31, 55, 85, 90, 93, 107, 116, 120, 121, 122, 195, 251, 267, 282, 284

ecology 97–101, 103, 105
elementary (school, teacher) 13, 30, 56, 59, 61, 67, 168, 170, 183, 210, 215, 259, 287
energy 2, 4, 46, 47, 86, 269
engineering 1–3, 11, 16, 21, 38, 47, 69, 70, 75, 85, 87, 88, 90, 109, 111, 129, 145, 157, 167, 181, 182, 194, 195, 197, 211–213, 227, 228, 230–232, 238, 243, 250, 252, 253, 258, 260, 264, 283, 286
environmental 30, 45, 52, 68, 91, 92, 100, 103, 115, 130, 135, 168, 282, 283, 288
equity 148, 151, 155, 156, 166
ethnic xxi, 14, 126–130, 133, 142
Every Student Succeeds Act 8, 39, 166

Fibonacci xii, 15, 227, 229, 234, 236–243

game 46, 61, 63, 77, 199, 209, 211, 251, 253, 254, 255
gender 1, 146, 171

higher education 10, 11, 87, 193, 195, 197, 200, 203, 205–207

identity xiv, xv, xviii, 1, 128, 201, 214, 288, 291
imagination 14, 38, 41, 42, 47, 55, 65, 70, 78, 93, 94, 108, 252
immersive 14, 16, 58, 107, 120, 167, 198, 203
inclusion 3, 7, 8, 12, 73, 88, 92, 94, 155, 190, 196, 265, 282, 286
inclusive learning 4, 15, 145, 167
industry xiv, 85, 86, 92, 94, 95, 132, 196–198, 203, 235
inquiry Learning 100, 102, 182, 217, 264, 269
interdisciplinary xvi, 4, 14, 16, 23, 38, 40, 44, 46, 48, 67, 73, 78, 79, 95, 97–99, 101, 102, 105, 156, 181, 193, 196, 198–201, 206, 207, 250, 253, 259, 267, 291

K–12 7–10, 22, 38, 87, 193, 195–198, 205–207, 211, 212, 227
kaleidocycle 249, 254, 255

lab 88, 89, 111–113, 115, 117, 118, 170, 173, 174, 185
literacy 1, 11, 15, 42, 45, 73, 89, 97, 126, 129, 142, 143, 165, 190, 205, 206, 210–215, 219

maker movement 2, 249, 251–253, 257
makerspace 112, 113, 121, 212, 281, 283
Makey Makey® XIII, 220, 221, 226, 271–274
mathematical 11, 15, 16, 53, 55, 76, 98, 155, 181, 182, 189, 190, 227, 229, 230, 232, 234–238, 240–243, 254, 255, 274, 282
multidisciplinary 24, 72, 198, 200, 250, 251
multimodal 212–215, 222, 223
multimodality 212, 214, 215, 219

narrative XII, 142, 215, 218, 219, 221, 288
natural science 14, 97-99, 101, 102, 147, 188, 189, 236
nature 41, 46, 52, 57, 64, 65, 100, 102, 105, 110, 112, 122, 162, 163, 168, 170, 174, 176, 194, 199, 200, 241, 253, 259

physics 11, 77, 91, 111, 168, 188, 189, 239, 259
play 43, 51–53, 56, 57, 59, 139, 182, 190, 212, 215–219, 222, 234, 253, 271, 273, 288
preservice teachers 161, 169, 170, 183, 215, 218, 269-276
professional development 40, 41, 71, 95, 182, 281, 282
programming 61–63, 89, 90, 113, 114, 116, 165, 200, 206, 211–213, 216, 217, 221
project-based 1, 3, 5, 10, 12, 25, 27, 30, 100–102, 105, 250, 251

robotic 12, 73, 85, 87, 121, 251, 253–257, 259, 283

Scratch XII, 213, 219–222, 226
ScratchJr. XII, 211, 216, 217, 222, 226
self-actualization XIX, 14, 51–54, 56–58
self-awareness 27, 29
self-expression 3, 54, 120, 122, 167, 194
self-management 25, 27, 29, 31
sensory 3, 42, 63, 147, 150, 154, 157
social awareness 27, 29, 31, 32
social emotional learning (SEL) 23, 27, 28, 31, 32
STEM to STEAM 7, 14, 55, 74, 78, 85, 86, 88, 89, 109, 267, 268, 277,
storymaking 210, 211, 214–216, 218–223
storytelling XV, 214

teacher training 3, 73, 74, 85, 206, 251, 268
tessellation 274-276
transdigital XII, 219–223, 226
transdisciplinary 10–12, 14, 15, 21–26, 30–32, 38, 40, 44, 48, 89, 212, 214, 227–229, 231–233, 243, 281, 291
transformation 3, 86–89, 174, 240, 241, 254

Universal Design for Learning (UDL) 155, 166–168, 176

virtual reality XVI, XIX, 2, 12, 15, 115, 118, 162, 193, 198, 209
visual culture XV, 1, 42
visualization XV, XXI, 14, 77, 109, 133, 183–185, 189

zoukei asobi (artistic play) 14, 51–53, 56, 59, 68

www.ingramcontent.com/pod-product-compliance
Lightning Source LLC
Chambersburg PA
CBHW031321230926
51848CB00001B/35